Cooking with Sea Vegetables

Cooking with Sea Vegetables

Sharon Ann Rhoads
with the assistance of Patricia Zunic

Illustrated by Sharon Ann Rhoads

AUTUMN PRESS

Published by *Autumn* 🜂 *Press,* Inc.
with editorial offices at
25 Dwight Street
Brookline, Massachusetts 02146

Distributed in the United States
by Random House, Inc., and in
Canada by Random House of Canada, Ltd.
Copyright © 1978 by Sharon Rhoads
All rights reserved.
Library of Congress Catalog Card Number: 78-51389
ISBN: 0-394-73635-4
Printed in the United States of America
Typeset at dnh, Cambridge, Massachusetts

Illustrations by Sharon Rhoads
Book design and typography by Beverly Stiskin

Table of Contents

Preface
Acknowledgements

I. Sea Vegetables: Past and Present

   The Bounty of the Sea 11
   Plants of a Thousand Uses 22
   Preparing Sea Vegetables 30

II. Cooking With Sea Vegetables

   Kombu and the Kelps 39
   Arame 54
   Wakame 61
   Nori 75
   Hijiki 91
   Agar 101
   Irish Moss 114
   Dulse 117

Sources of Sea Vegetables 127
Nutritional Value of Sea Vegetables 128
Bibliography 129
Recipe Index 132

Preface

This book began during a 1977 New Year's visit to the house of some Japanese friends in New York, to enjoy a feast of traditional Japanese festive dishes. According to custom, as many as thirty different delicacies were arranged like a decorative collage in beautifully lacquered stacked boxes. Almost half of the dishes contained sea vegetables, which are considered particularly auspicious. All were delicious. Suddenly I realized that there was no book in English that really drew upon the highly developed cuisines of the Far East to present the time and taste-tested ways of preparing these sea herbs.

Once the research began, it soon became apparent that sea vegetables offer far more than tasty substitutes for or adjuncts to land vegetables. Their marvelous nutritional and medicinal properties are only beginning to be explored; they are indispensable to a multitude of industries, and their probable future role in methane gas production may make them a major benefactor in the search for alternative energy sources. Sea vegetables truly represent the bounty of the sea.

In preparing the recipes, the already existing standard sea vegetable dishes of the British Isles, China, Korea and Japan were first explored, then adapted to American tastes and American kitchens. Gradually, experimentation enlarged to cull from cuisines around the world. Many of the recipes presented here are original, or original adaptations.

Every one of the recipes in this book uses strictly natural, wholly vegetarian foods. There are no meat, eggs, or dairy products of any kind. With kelp powder on the shelf, even sea salt is rarely needed. Sea vegetables do, of course, go perfectly with nonvegetarian foods, enhancing their flavor just as they enhance the flavor of purely vegetable dishes. The cook should feel confident to include sea vegetables in any favorite dish.

Our work with sea vegetables has strongly reinforced our conviction that modern living has removed us from being in touch with the land and with the needs of our bodies, but that the food we eat can be our ticket back. It can help us to restore the balance that has been upset. Not only can food be the best medicine (particularly in the vast number of afflictions resulting from mineral or other nutritional lacks), it can also offer us an opportunity to improve our general health. Sea vegetables can be enjoyed by people on any kind of restricted diet; they are ideal remedial foods. Even those in the best of health, however, can benefit by daily consumption of these versatile, savory gifts from the sea.

By suggesting a satisfying, all-inclusive range of purely vegetarian dishes, we have tried to show that food that is good for you can also taste good. (We can certainly vouch for that fact, and so will all our friends and relations!) Cooking with natural foods is always exciting. Even when the kitchen is in a city apartment, the food can still be prepared with loving care and gratitude for the natural abundance that it represents. There is a great joy in knowing that the food we are preparing will nourish body and spirit. By cultivating a consciousness of the bounty of the land and sea, we reestablish our link with the greater natural world of which we are an important part.

Acknowledgments

My first and greatest debt of gratitude goes, of course, to Patricia Zunic, who helped to develop the entire book from the beginning, including the initial research and planning. One-third of the recipes are hers, and her extensive herbal and culinary knowledge were drawn upon throughout.

To Jeff Maron, whose generous support in every way made this book possible: my deepest thanks. I am also indebted to Irving and Jeanne Maron, Catherine and Carl Lobell, Aleta St. James, and many others too numerous to mention who provided continuing help and encouragement and were always willing to sit down to a "seaweed meal," no matter how wild the experiment.

I wish to express my gratitude to countless Japanese friends who have guided me in an aesthetic appreciation for the delights of oriental cooking. Without the assistance of Hiroko Takahashi, my "Tokyo correspondent," this project might still be in the planning stages. A vote of sincere thanks goes also to Erewhon and the American Museum of Natural History; and especially to Shirley Corvo, Susan Willis, Deborah Balmuth, Sandy MacDonald, and Beverly and Nahum Stiskin for their patient and invaluable assistance.

—Sharon Rhoads

I. Sea Vegetables: Past and Present

"Bounty of the Land; Bounty of the Sea"

The Bounty of the Sea

The Secret of Eternal Youth

Finding the Elixir of Immortality was an obsession of the emperor Shih Huang-ti, who built the Great Wall of China in the third century B.C. Prophecies foretold that the substance bestowing the eternal youth of the Taoist immortals was to be found on "an island in the Eastern Sea," and Shih Huang-ti dispatched many emissaries to the east in search of the precious material. One of these, Hata of Ch'in, is said to have set off for the islands of Japan, located in the Pacific to the east of China, and there to have come upon the secret. Hata returned to China once, but he then went back to Japan without divulging the secret, and the Chinese never saw or heard from him again. Many Japanese say that Hata discovered the secret of immortality in sea vegetables.

There is, in fact, a good deal of truth in the tradition that attributes such great powers to these humble marine plants. Although sea vegetables may not offer an antidote for the natural transition known as dying, researchers have discovered that they do have many properties that work to prolong life, to counteract the effects of aging, and to make the lives thus lengthened far healthier and happier.

In celebrating the New Year, Okinawans traditionally sing:

In the fresh new year
We offer up the charcoal and the kombu
From the spirit to the body,
All becomes young.

Charcoal represents the bounties of the land. In other places in Japan, other symbols are used (most commonly, moon-shaped rice cakes with dried persimmons or a sprig of an evergreen bush), but everywhere the bounties of the sea are represented by one or more fronds of kombu, a popular sea vegetable belonging to the group known loosely as "kelps." The range of uses of this plant more than justifies its place of honor atop the festive rice cakes; and, as the song suggests, the Japanese were moved to place it there for its properties of rejuvenation as well as for its "evergreen" qualities (the symbolism that moves us to put up a Christmas tree to carry us forth into the new year).

Recently, Professor S. Kondo undertook an ambitious research project to determine which regions of Japan sustain people to the ripest old ages and why. He surveyed every district of Japan and found that people grow older and look younger in those areas where sea vegetables are eaten abundantly. The famed women pearl divers of the Isé Peninsula, who dive to considerable depths virtually naked, summer and winter, well past their seventieth birthdays, told him that they are careful to eat lots of wakame, hijiki, and nori every day. Major causes of death in later years (such as cerebral hemorrhages and high blood pressure) are notably rare in these areas.

Good health is the best preventive against illness; cultures the world over have understood that sea vegetables not only provide an excellent food with an almost endless variety of palatal pleasures but serve equally well as wonderfully effective agents in disease prevention and healing.

Beauty is another gift that sea vegetables bring, for vibrant health imparts radiant beauty. The Japanese hold that wakame, hijiki, and arame consumed daily will ensure thick, glossy hair, a clear complexion, and soft, pliant, and wrinkle-free skin.

The Welsh hold that sea vegetables have "magical powers." And they do indeed. In *Common and Uncommon Uses of Herbs for Healthful Living* (Parker Publications, 1969) Richard Lucas recounts the following story: In the winter of 1967–68, an army of hares descended upon the Kamchatka Peninsula in far eastern Siberia. Ignoring barking dogs and astonished humans, they marched through towns and across the frozen land in a straight line toward the sea. As soon as they reached the beach, they devoured sea kale that had been swept ashore by the tide. The hares, it was deduced, were mineral-starved.

Sea plants contain ten to twenty times the minerals of land plants, and animals seem to know this instinctively. The coast-dwelling Irish and Scots know that their livestock always graze on the twisted strands of sea vegetables strewn along the beachheads in preference to thick green grassy fields. The British Islanders also affirm that these "sea green-fed" animals are sleeker and healthier and give better milk than their land-locked counterparts. The same qualities that make the sea vegetables such super-fodder have led to their wide-scale use as fertilizers, soil-builders, and soil-restorers. They have even been used to create soil where there was none!

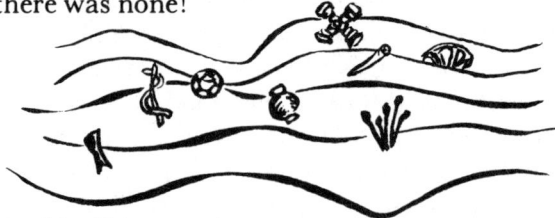

Fossil algae include the oldest living organisms

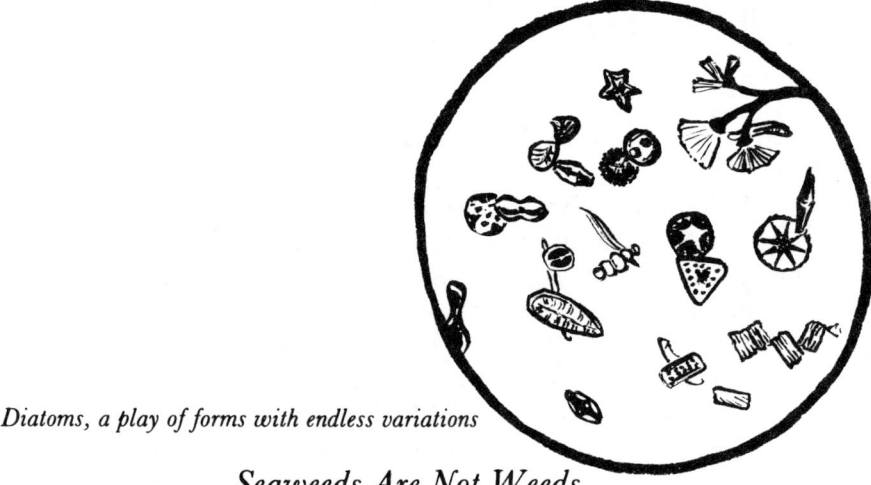

Diatoms, a play of forms with endless variations

Seaweeds Are Not Weeds

The term *seaweeds* usually refers to the macroscopic varieties of marine algae. Algae are the first classification in the plant kingdom: they comprise the first seven phyla (or eight, depending on the source of classification) in the plant kingdom and encompass an astoundingly diverse spectrum of living things, including the longest plants in the entire kingdom as well as the minute one-celled organisms visible only through an electron microscope. Giant kelps stretch their fronds for hundreds of feet through coastal waters. As a visual banquet, the microscopic diatoms, whose silica shells present an endless variety of sculptural designs, are rivaled only by snowflakes in the natural world. Along with the other one-celled algae and tiny marine organisms, they constitute the floating population known as plankton. Thus, algae represent the lowest part of the food chain, supporting fish and crustaceans, which may be food for even larger fish and crustaceans ultimately eaten by humans.

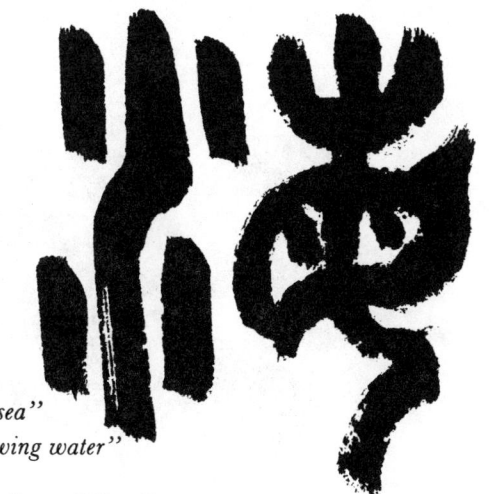

The ancient Chinese character for "the sea" depicts a full-breasted mother beside "flowing water"

The Sea, The Source

In myth, all things arise from and return to the sea. For centuries the Hindus have spoken with reverence of the great "shoreless sea," the cosmic ocean where Vishnu resides. The source of all things in the universe, this sea represents a crystallization of *maya*: innumerable worlds rise from these waters and settle back into them, in a beautiful, never-ending succession.

Almost every culture carries in its collective memory a recognition of the importance of the sea in the creation of life. When the Chinese conceived their character for the sea, they envisioned the primordial Mother with full, milk-rich breasts and combined this figure with the radical used for phenomena or concepts concerned with water (see figure). They assigned the phonetic reading *hai* to the character because it means, "green and wide in expanse"; to them, the sea was not only a mother but a garden.

The ancient Sumerians numbered among their deities the goddess Nammu, "the mother who gave birth to heaven and earth"; her name is written with the character for "sea." Many cultures, like the Jicarilla Apaches, acknowledge in their myths a time before creation when all was "darkness, water, and cyclone."

Modern science has lent credence to the myths: paleontologists have confirmed that, indeed, we must look to the ocean for the origin of living things. As convincing as the fossil evidence may be, we carry the greatest proof of our sea beginnings in our own bodies. "Ontogeny recapitulates phylogeny": the evolution of the human race is mirrored in the development of the individual human body, which begins in the womb's saline solution, the amniotic fluid. The constitution of the blood that nourishes and cleanses our bodies is nearly identical with that of sea water; so close are they, in fact, that saline solutions have been used successfully for transfusions when blood plasma was unavailable.

The amount of salts in the sea has increased as the earth has aged, and the blood of land vertebrates has a denser salt content than that of marine vertebrates. Thus, science has been given a clue concerning the order of evolutionary development: the denser the blood salt content, the newer the species.

The newest species, Homo sapiens, requires certain elements for the maintenance of its metabolism: some of the primary ones are iodine, copper, calcium, phosphorus, manganese, sodium, potassium, magnesium, chlorine, and sulfur. Moreover, iron, copper, manganese, zinc, and aluminum ensure rich blood and act as catalysts in body processes. All of these minerals are abundantly accessible in sea water (one of them, iodine, is difficult to obtain in sufficient quantities from any

other source). Sea water contains approximately 3.5 percent dissolved inorganic compounds (minerals). It is estimated that 1 cubic mile of ocean water carries 200 million tons of such chemical compounds. And sea vegetables contain them in abundance.

The minerals in the soil are constantly being leached out and carried back to the sea. There they are absorbed by the algae, which then hold them in suspension, an open invitation to consumption by sea urchins, fish, or humans. In solution, the elements present in sea water can be more readily assimilated by the human body. Although the concentration of elements is generally constant among all algae, it may vary slightly from plant to plant. (Sea water itself maintains almost constant salinity below the tide lines; in the more open sea, water evaporates at a higher rate, increasing the concentration of salts available to the algae.) Algae do not, however, absorb pollutants along with the other elements taken in in such concentrations.

Where the level of pollution is high, they fail to grow (as is evidenced by the shockingly reduced nori crops in Japan's polluted sea waters). In fact, elements in certain sea vegetables actually remove radioactive and toxic metal waste from the body.

Sea vegetables offer many elements in up to several thousand times their concentration in sea water. Iodine, for example, is an element normally found only in sea water, and some sea vegetables may accumulate up to 20,000 times the density of iodine in sea water. Therefore, the major iodine-deficiency disease, goiter, is unknown among populations that eat sea plants. Calcium is another important element present in high concentration in sea vegetables. The human body contains an entire kilogram of calcium, which is largely used as an important constituent of bones and teeth. Calcium also works to calm the nerves. A lack of it interferes with the blood's clotting ability and the flexibility of muscles and sinews; a severe shortage can even result in nervous tetany. Furthermore, since the body consumes

Evolution of the Chinese character for "the sea"

calcium daily, it is important to include high calcium foods like sea vegetables in the daily diet.

Increasing numbers of diseases are now ascribed to deficiency of some element or vitamin; symptoms can be greatly alleviated through massive dosages. (Thus, biochemical tissue salts are coming into wider usage.) Also, doctors have begun to prescribe kelp for a wide range of mineral-related afflictions including arthritis, rheumatism, obesity, high blood pressure, and thyroid problems.

The nutrient blessings bestowed by sea vegetables, moreover, are not limited to the inorganic salts. Why is cod liver oil so high in vitamins? When biochemists analyzed the fish oil in order to answer this question, they found that the vitamins came from the smelt eaten by the cod, and the smelt got them by eating diatoms and their algal relatives. The algae are the original producers of A and D, which they manufacture in quantity. Being oil-soluble, these vitamins are found in tiny oil droplets in the protoplasm of the algae. Some researchers have recently suggested that algal vitamins are manufactured by bacteria living on or around the plants. In fact, a cod liver alternative has been developed by combining roasted bladderwrack (*Fucus vesiculosis*) with a bland oil.

Certain sea vegetables are absolute powerhouses of vitamins. Asakusa-nori has ten times the vitamin A of bell peppers and fifty times as much as tomatoes. Nori and wakame offer ten times the niacin of spinach and at least as much vitamin C as tomatoes; they are high in B_1, B_2, pantothenic acid, and B_6 complex. From dulse one can get biotin, D, E, and B_{12}, the anti-anemia vitamin otherwise lacking in many vegetarian diets.

Even protein is available through sea vegetables: 3-1/2 ounces of dulse a day constitute an adequate protein source. Granted, that's a lot of dulse, but still, it is no wonder that in many parts of the world sea vegetables have proven to be lifesavers in times of famine.

In addition, sea vegetables impart an ineffable quality beyond the elements necessary to life. Scientists have "created" artificial sea water in their laboratories with all the elements present in the ratios in which they are found in natural sea water. But when sea urchin eggs are placed in this environment, they will not hatch—unless other living creatures are present. Scientists speculate that the other organisms add something "organic" to the water, though they cannot define that organic substance using the periodical table of elements. It takes life to support life; thus, sea vegetables are a treasure because they provide vitamins as part of a whole, living organism, and they can therefore be more readily assimilated by the living cells of our bodies.

Simplicity: The Key to Nature's Treasure

The sea vegetables' nutritive properties are directly related to their simple biophysical character. Algal reproduction takes place asexually (through simple cell division or regeneration from fragments) and sexually (by means of spores). When conditions are favorable to the sea vegetable population, the spores germinate to produce new plants. Pioneering in the development of life forms and remaining at the bottom of the food chain, the algae have maintained an undifferentiated structure; that is, even the largest of the sea vegetables are composed throughout of a single, unspecialized type of cell. (It was left to the land plants to develop the complex system of specialized cell groups that make up roots, stems, and leaves.) Many microscopic algae and even some macroscopic algae float free (one such is *Sargassum*, whose namesake, the Sargasso Sea, is famous as a hazard to sailing ships on account of the vast jungles of bobbing fronds). To fix themselves to something stable in order to enjoy the benefits of a certain shoreline or offshore depth, other sea vegetables developed disc-shaped or tendril-like "holdfasts," aggregates of cells at the base of the plant that do just what the name suggests, holding fast to rocks, seashells, wharfs, pilings, or other algae. The rest of the plant consists of a long "stipe" (or stem) which separates into fronds. These plants are so much "of a piece" that it is sometimes difficult to distinguish where the stipe ends and the fronds begin.

The simplicity of their structure enables sea vegetables to absorb nutrients through the entire surface of the plant, instead of only through their roots, in the way that land plants take up food. Floating in the moving waters, they had no need to develop a rigid, thick cellulose support; thus, they are more succulent and require less cooking than most land vegetables.

Moreover, the sea does not impose as many limitations on its inhabitants as the soil does. The temperatures in any zone are fairly consistent year round. Nutrient salts may be present in slightly higher concentrations in winter than in early summer, but they are always available. Oil spills and other severe pollution aside, the ocean waters are always in motion and never stagnant. Furthermore, they are constantly being replenished by rivers carrying their rich loads from the continents. Barring any foolish mistakes on the part of humanity, we need never fear that the pastures of the sea will become parched or sterile.

All of these factors add to the attractiveness of sea vegetables as a food source. Every part offers the same plentiful store of minerals and vitamins as any other. They are easy to harvest and, if harvested in suitable amounts, require a minimum of encouragement to replace themselves. Every known species of macroscopic alga is edible. Nature has laid virtual treasure chests at our feet along nearly every coast on the globe; we need only act sensibly in taking advantage of these gifts.

Light & Color = Good Food

Botanists classify the sea vegetables by color: the reds, or Rhodophyta; the browns, or Phaeophyta; the greens or Chlorophyta; the blue-greens, or Cyanophyta. Because of the color similarities the latter often are subsumed under the Chlorophyta, and occasionally taxonomists will distinguish a separate group, the Xanthophyta, or yellow-greens. The pigmentation that provides the basis for this classification is related to the spectrum of light to which the plants have access for photosynthesis.

The Rhodophyta, the most light-receptive sea vegetables, can be found at the furthest depths, 200–600 meters, where they absorb the shortest frequencies of sunlight, the blue and ultraviolet waves. The sea vegetables pick up these waves with their red and blue pigments, known respectively as phycoerythrin and phycocyanin. This group, which evidences other colors in addition to red, also flourishes in rock pools on shores or below the low tide mark. There are as many red species as greens and browns combined. The reds most commonly used in cooking include dulse, purple nori or laver, Irish moss, and agar.

The yellow-brown pigment fucoxanthin colors the phaeophyta, including the wracks and kelps. These brown plants encompass the largest of the sea vegetables (and thus the longest of all plants). Phaeophytes prefer cooler climes, such as the shores of New England, Canada, Hokkaido and other areas of northern Japan, and the British Isles. They live at medium depths down to 50 feet, where they can absorb medium, green wavelengths of sunlight.

The Chlorophyta provided the main line of evolution from water to soil, culminating in the higher land plants. Absorbing the long, red wavelengths from the surface of the waters, where they are exposed to the strongest light, they store food as starch just as the higher plants do. The Cyanophyta, often subsumed under the chlorophyta, contain the blue pigment phycocyanin, along with the pigments xanthophylls and carotene. This group includes green nori and sea lettuce.

The reds and greens are said to have the highest concentrations of vitamins B_1, B_{12}, and pantothenic, folic, and folinic acids. The possible correlation between the pigment, light exposure, and the distribution of nutrients in each class has yet to be established, but further research in this area will undoubtedly accompany the growing interest in sea vegetables as food.

The Myriad Sea Worlds

The seashore and the tides that wash over it create a variety of environments, each with its own typical associations of animal and plant denizens, all interacting to support the whole. These biological "zones" include the *littoral* (which extends broadly from the high tide line to the edge of the continental shelf), the *intertidal,* and the *low tidal,* with a number of intermediate bands in between. In addition, the rock pools and emptying salt marshes afford specific sets of conditions that are attractive to certain creatures and plants. The brown sea vegetables, for example, commonly reside just below the low tide zone. They are strong enough to thrive in the turbulence of the crashing waves and well adapted to making full use of the lesser sunlight available there. The red sea vegetables live below the browns in less agitated waters and are thus permitted finer, more delicate

Harvesting Ascophyllum off the coast of Norway

branching. Green algae are most versatile. They are found anywhere and everywhere, between tide marks and rock pools or coloring the surface of the waters with filmy or filamentous masses that swell and dip beneath the waves.

The ocean is not one but a myriad of worlds. The beauty and complexity of its forms inspire poetry, and the intricacy of its interrelationships and the great harmony of the whole excite awe. The rockweeds, for example, furnish countless organisms with a jungle home that exhibits every bit as much activity and life drama as the jungles of Africa. Turbulent waves constantly pound at the rockweed fronds, breaking off bits and pieces and releasing them into the ocean currents to feed nitrogen-fixing bacteria and plankton—which in turn support the crustaceans, hydra, fish, and many other species. The elements taken up by these animals ultimately find their way back into the sea water, to be absorbed by the sea vegetables once more.

With the exciting, often spectacular, leaps that marine science has made in recent years, we have gained a greater understanding of subtleties of the oceanic environment. Sea vegetables offer a generous food supply, but we must govern their use with care, so that the bounty of the ocean may be available to our children's children and theirs, ad infinitum. Pollution in the Seto Inland Sea has already seriously endangered the Japanese nori crop, signalling a warning around the world that when we upset the balance of nature, we endanger the seas and everything that depends upon them, including ourselves.

Gathering the Crop

The sea is generous with its presents. Waves roused by seasonal gales wash great quantities of driftweed up on the beaches, where it is easy for harvesters to rake up the crop and cart it off. Even in these days of mechanization, sea vegetables are still harvested mostly by hand or from small boats. In Ireland, the driftweeds for sea vegetable meal are hand-mown with sickles; in Spain divers collect the crop under water. Around Cape Cod, harvesters in small boats catch up Irish moss with an 18-foot rake with closely set, 6-inch teeth, yank up the plants, and wipe the rake off in the boat. Their "catch" is sun-dried for packaging on shore or inland. Despite the laborious methods of sea vegetable harvesting, the Chinese proved the viability of labor-intensive methods by training 20,000 people from Fukien Province in the techniques of kelp cultivation before the 1960s.

Most sea vegetable harvesting enterprises are small, family-run businesses. Some machines have been introduced. In Scotland tractors are outfitted with fork lifts to heave the driftweed into a wagon—which may well be horse-drawn. Atlantic Laboratories in Boothbay, Connecticut, founded in 1971, originally brought in its crop by hand; now the sea vegetables are mown with power-driven sickle bars mounted on boats.

Sea-vegetable harvesting in Scotland

The kelp industry involves the largest-scale, most highly mechanized harvesting, particularly on the California coast. Mowing is accomplished by means of big barges (the largest cuts a swath 18 feet wide and pulls in 200 tons in only 6 hours) equipped with vertical blades on the sides of a single horizontal blade. The blades are dipped 3 feet into the water so as to shear off only the tops (to ensure future crops, the government has regulated the amount that can be cut off). In Alaska, where as much as 4.9 kilograms (almost 11 pounds) of *Macrocystis* can be harvested per square meter, mechanical harvesting machines mow the stipes and a chain elevator lifts them onto the ship.

Several difficulties have inhibited growth of the industry in many areas of the world, however: many species like to intertwine or grow on one another, making it difficult to collect the pure crop. Many sea vegetables prefer surf-battered, rocky coasts which offer no access by boat. Since much of the work must be done by hand, labor costs are prohibitive. Driftweed gatherers are dependent on the whims of the weather to furnish their crop.

Nonetheless, the Food and Agriculture Organization of the United Nations projects a tenfold increase in the world market for sea vegetables, and as demand increases, industry will surely find ways to meet these problems, although sea vegetable gathering may have to remain largely the work of small, independent companies or individuals.

Cultivating the Sea Garden

Like island dwellers the world over, Hawaiians have long cultivated the sea. Their staples included, in addition to fish and taro, over seventy species of sea vegetables called *limu*. Their methods of cultivation were simple: encouraging the growth of choice varieties in coves and sheltered places by weeding out unwanted algae competitors. When a new chief ascended to the throne and moved his capital, his garden would be transferred with him and reestablished in an auspicious location.

Since sea vegetables reproduce mainly by regeneration from fragments broken off parent plants and from spores, for a long time no one developed cultivation methods beyond seeding protected areas with rocks (to which spores would attach) and eliminating competitors and predators. The Japanese were the first to introduce artificial sea vegetable cultivation, in the production of nori.

Starting in 1736, bundles of oak brushwood or bamboo (called *hibi*) were sunk into Tokyo Bay to a depth of 3–5 meters; after the nori spores attached and grew into good-sized plants, gatherers in small boats raised the *hibi* and collected the nori in small baskets. In the 1930s horizontal nets replaced the vertical bamboo. The horizontal net system is greatly superior, because it allows sea farmers to adjust the height of the nori population to suit changing conditions of light, temperature, currents, et cetera.

In 1949 an additional stage in *Porphyra tenera* (nori) reproduction was discovered as a result of advances in nori cultivation. Originally thought to be a different species altogether, the nori sporophyte (the spore-bearing generation of the plant) was known as *Conchocelis* because it grows in mollusc shells until the shorter days and lower temperatures of September trigger their release to float freely in the water. When the connection between nori and *Conchocelis* was fully understood, it enabled nori farmers to influence conchospore release by cultivating *Conchocelis* on oyster shells, then lowering water temperature. Culture nets are placed in estuaries that are high in nutrient content and then they are moved to the open sea. Careful watch is kept to avert infections, pollution, or consumption by the number one sea vegetable enemy—sea urchins.

Another recent Japanese innovation in nori cultivation involves creating an artificial sea floor with a network of nets or bamboo blinds. *Porphyra* nori is now Japan's principal sea vegetable crop, followed by *Laminaria* (kombu) and *Undaria* (wakame).

The Far Eastern countries presently cultivate the latter species through four major methods: planting large stones or concrete blocks on the ocean floor, cultivation on ropes, digging out flat reefs when exposed at low tide, and dynamite blowing (clearing rock bottoms with dynamite or creating new strata by blowing shelves in mountains at proper depths). Along the Japanese coast kelps attached to rocks take two years to mature, but after only one year they can be removed and transported and will still grow; thus, plants nurtured in nutrient-rich areas can be "transplanted," allowing for consecutive seeding of crops.

Plant growth is affected by a number of conditions, including temperature, salinity, and nutrient content of the sea water. The China coast, for example, is mostly too low in salinity and nutrient value, and too warm, to spawn algae populations. Therefore, the Chinese and Koreans have developed inland "hatcheries"; the sporophytes are transported to the coast. To compensate for the nutrient deficiency of the Yellow Sea, the Chinese devised porous cylinders filled with nitrates, which they drop from rafts (one of the beauties of algae lies in their ability to assimilate supplements of inorganic earth elements). The Chinese also intend to introduce nitrogen-fixing bacteria and blue-green algae into the deficient waters.

The benefits of cultivating sea vegetables, however, far outweigh the difficulties it presents: The Chinese have found that the laborer can manage 2 hectares of sea floor at a cost per year of only $400. The value of production for those 2 hectares in one year amounts to $8,000. For the enterprising individual, sea vegetables may well represent the crop of the future.

From Sun to Cellophane

The Hawaiians used more than one hundred words to refer to the various kinds of *limu*. They distinguished between "one-day *limu*" (which had to be eaten within a day of harvesting) and those which, dried or fermented, can be eaten as much as a year later.

The methods of preserving sea vegetables have changed very little since ancient times. Most are draped or laid out to dry in the sun—or, today, carried inland by truck to be dried by machine. In Japan nori is processed in the traditional manner: it is chopped fine, washed in fresh water, and set out to sun-dry like sheets of paper, on wooden frames. On the West Coast of the United States, kelp is carted to the factory, where it is cut, macerated, and dried by a mechanized process. Further processing varies with the sea vegetable. Kelp is often powdered as a spice or compacted into tablets as a nutritional supplement. The sea vegetables used for their gelling properties require the most complex treatment. Agar (*kanten*) was traditionally rinsed in fresh water, sun-bleached, re-rinsed, boiled for several hours in large vats, constantly stirred, filtered, freeze-dried, thawed, then cleaned several times and dried. Today extraction is achieved in autoclaves under pressure; then the agar is decolorized, deodorized, filtered through activated charcoal under pressure and evaporated under reduced pressure; it is finally frozen and thawed. Extraction of carrageenin from Irish moss is less demanding. After several fresh water baths, the *Chondrus* fronds are boiled for several hours, and soluble matter is separated out in a centrifuge; then the extract is partially dried in a vacuum and finally rotary-dried.

Increasing amounts of wakame, arame, hijiki, kombu, and nori are being sent to the United States, in cellophane packets and in bulk. Health food stores carry them, along with dulse and Irish moss from Newfoundland, Nova Scotia, Maine, and Massachusetts.

Domestic enterprises are burgeoning, most of them founded and managed by people who have themselves experienced the benefits of seaweed consumption.

One such individual, concert pianist Joseph V. Wachter,

came to America from Vienna at the height of the Alaskan gold rush. He was forced to abandon his career after contracting tuberculosis. By adopting the indigenous Northwest Coast diet, with its emphasis on sea vegetables, he not only regained his health but enjoyed an extraordinary robustness. After giving a concert at the San Francisco World's Fair, he walked to New York City and back. Convinced that he owed his newfound vitality to sea vegetables, he traveled around the South Pacific from 1920 to 1930, researching the native use of sea vegetables as food. In 1933 he established the Organic Sea Products Corporation in Burlingame in the San Francisco Bay Area, branching out from foods to include cosmetics, vitamins, and finally fertilizers. His enterprise, carried on by his family, is thriving: it boasts a new library and museum.

On the East Coast, enterprises like that of Shepard and Linnette Erhart in Franklin, Maine, are finding that they have more orders for wakame, dulse, kombu, and kelp than they can handle. Ideally, more American coastal farmers will turn to sea vegetable crops and benefit from the processing methods already highly developed in the East.

Plants of a Thousand Uses

In the inlet at Mama
in Katsushika, like dark tresses
it wisps and it flows,
and I dream of the mythical Tekona[1]
as she gathers the precious sea grass

It is as though
garments in scarlet are mirrored
in the Okami River
while young girls wade in the shallows
plucking the nori strands

—The *Manyoshu*

The image of lovely young girls reaching down gracefully to gather gifts from the ocean was very nostalgic for the Japanese of 1,600 years ago—and familiar as well. Many songs and poems in the *Manyoshu* ("Collection of Ten Thousand Leaves"), an eighth-century anthology, attest to the place that sea vegetables have occupied in people's daily lives for centuries.

Although they are the major producers and consumers of sea vegetables today, the Japanese are by no means unusual in their extended use of marine algae. The Emperor Shen-neng of China supposedly used sea plants as both food and medicine in 3000 B.C. A domestic poem in the *Chinese Book of Songs*, a collection of poetry dating from 800 to 600 B.C., depicts a housewife cooking sea vegetables, and the *Erh Ya*, "the oldest known Chinese encyclopedia," from before 300 B.C., records twelve species. Many centuries B.C. the Koreans sent laver, kelp, and

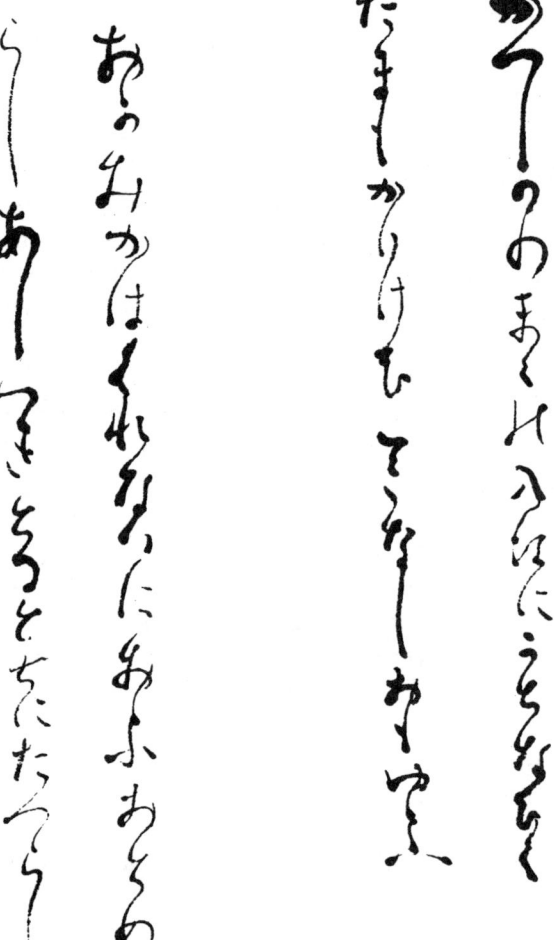

1. Tekona was a legendary beauty whose charms were so great that a swarm of suitors finally drove her to despair, and she drowned herself in a river.

agar to the Imperial Court of China, where they were prized for their medicinal properties. All the major sea vegetables consumed in Japan today were registered as part of the annual tribute to the Court in the eighth century A.D.

For a long time sea vegetables were used to collect salt. Species thrown up on shore were gathered, piled, and dried; sea water was poured over them; they were redried many times, until they were saturated with sea salt; they were then burned, and the ashes were boiled in a pot of fresh water, which evaporated, leaving the salt.

Since sea vegetables were so important in daily life, they were naturally incorporated in rites of fertility and thanksgiving. In Okinawa, one of the southernmost Japanese islands, a spring festival was held on the third day of the third month (by the old lunar calendar) to ward off misfortune. The women would fix stacked boxes of luncheon treats (including lots of sea vegetables, of course) and go to the beach to gather *mozuku* (*Nemacystis decipiens*), tucking up their kimonos and wading knee-deep into the water to gather the slippery plants. Mozuku was—and still is—served with vinegar and ginger slivers as a favorite accompaniment for sake, or as a tasty accent for baked fish or shrimp.

The Japanese, Koreans, and Chinese were not the only peoples to hold the sea vegetables in such high esteem. Pythagoras mentioned edible varieties in his writings, and the Aztecs cultivated algae. *Limu* was a staple for the native Hawaiians, who ate over seventy species until mainland eating habits replaced tradition in this century. The Vikings lived on sea plants when faced with starvation in the ninth century. The British Islanders' native nori (*Porphyra*), "laver" (Latin for "water plant") has been in use since Roman times. Dulse and *Porphyra* were once eaten in Scotland and Ireland fresh, dried, or cooked like spinach. The Irish even chewed dulse like tobacco. The southern Welsh still eat and export "laverbread"; the laver is boiled, salted, drained and minced, then warmed in fat or made into cakes and coated with oatmeal. Sea vegetables are used instead of salt in some areas of Britain to preserve cheeses. *Porphyra* (called *karengo* in New Zealand) served Maori soldiers well in the Second World War when they were sent to the arid Middle East; they chewed it on long walks and never got thirsty.

The formidable stipes of the *Laminaria* have also served as walking sticks, knife handles, and fishing rods and line. If the stipe is straightened when fresh and allowed to harden, it dries very hard—hard enough to be tied to harrow teeth for harrowing, as it is used in St. Kilda. Sea plants have played a vital role in the lives of almost every people throughout the world—Southeast Asians, the Japanese, the Indians of Siberia, Tasmanian aborigines, South Sea Islanders, Eskimos, and the natives of coastal areas in Europe and the Americas. Remains of sea vegetable banquets have even been found in the garbage dumps of Stone Age South African dwellings.

Good Food, Good Medicine

The sea vegetables have often been advanced as purveyors of healing, not only in rite but in empirical medicine—and not only in the past, but up to the present. One of their greatest assets as a medicine is that, being a food, they do not cause any of the harmful side effects which often accompany the use of drugs; in fact, they can even work to assuage them.

Traditionally, when a Hawaiian shaman wished to do an

incantation to drive away sickness, he would gather up a species of *Sargassum,* make a lei out of it, and ingest a small part of the plant. With the lei in hand, he would stand at the edge of the ocean, offering a prayer of penitence—and promise of reform to follow on the healing—on behalf of the sufferer. Then he would stride into the sea. With the water licking about his body, he would place the lei on his head or neck, eat a bit more, and then throw it into the ocean, while reciting a prayer requesting the revitalizing energy of the sea. He would not look back, speak, or gesture to anyone until the rite had been accomplished, for his contact with the powers of the sea, embodied in the sea plant, was not to be broken. Hawaiians also boiled infusions of *Hypnea nidifica* for stomachaches, dried nori for boils or used it as a poultice, or cooked *Centroceras clavulatum* for use as a cathartic.

In the tenth century the Japanese compiled a two-volume encyclopedia of medicinal plants that had been proven through use in Chinese medicine for many centuries. The *Honso Wamyo* pays homage to the medicinal efficacy of sea vegetables in the treatment of a long list of ailments, noting, for example, that "mozuku is good for warts and swellings; nori for warts and rickets; arame for feminine disorders and mouth afflictions; aonori for warts, hemorrhoids, stomach ailments . . . ."

In the West, Richard Martin prescribed raw dulse as an antiscorbutic in 1695; earlier, George Camden indicated laver for "all distempers of the liver and spleen . . . Dr. Frank Owen found relief from it in the acutest fits of the stone." The ancient Incas gave massages with sea algae and valerian leaf ointment at the first signs of illness.

In modern-day Korea a diet already rich in sea vegetables is supplemented with kombu for circulatory or feminine ailments, as well as during pregnancy and after birth. In Ireland anyone with a "whitlow" (a painful, pus-producing inflammation) on the hand is advised to stuff the hand into a pile of rotting Irish moss (*Chondrus crispus*) and to leave it there "as long as patience will allow"; when pulled out, the hand will be cured. Dulse is used in Skye to induce a sweat during fever. In Iceland it is recommended as a remedy for seasickness.

Modern scientific research has not only confirmed the efficacy of traditional remedies using seaweed but has introduced new cures.

Seaweeds in First Aid

A long folk history of kelp ointments and linaments for cuts, stings, sprains, and bruises is supported by the use of alginates as hemostatic powders and surgical dressings. The alginates increase the rate of healing without inducing a toxic reaction or antigen response in the body. Equally important is the fact that alginates do not interfere with concurrent treatments by drugs, such as thrombin, penicillin, or the sulpha drugs. Even a minor injury, such as a bitten tongue or cheek, can be quickly assuaged by chewing a little nori and placing it over the wound; this stops the bleeding and diminishes the pain. The algal mucilages provide ideal airtight gel "bandages" for burns, relieving pain by holding moisture in the damaged tissues and keeping oxygen out. These same mucilages in sodium alginate produce soothing gel-like "rafts" floating on top of stomach contents, to keep the acid down and prevent heartburn (see *Lancet,* January 26, 1974). Applied by doctors in a variety of skin ailments, sea plants are a common component of many skin creams and health shampoos now on the market.

Increasingly, modern medical research is tracing many of humanity's afflictions to nutritional lacks or imbalances. In many cases sea vegetables can help to restore such nutritional balances. For example, the Scripps Oceanographic Institute found that a sea vegetable diet prevents hay fever. Always, proper daily nutrition is truly preventive medicine.

Sea vegetables comprise 20–30 percent protein. Readily digestible and equipped with a full complement of vitamins and minerals, they are among the few complete protein sources in the vegetable kingdom. ("Complete protein" is protein containing all of the eight essential amino acids; if any of these amino acids is missing, the protein cannot be assimilated.) They contain: 1–9 percent fats, present as the fat-soluble vitamins A, D, E, and K; essential fatty acids; and lecithin and necessary sterols, such as cholesterol[2] and ergosterol, both of which, in the presence of sunlight, are converted by the body into vitamin D. The Japanese have long used aonori and another member of the Chlorophyta (green) family, *Monostroma,* to reduce cholesterol in blood plasma.

Sea vegetables contain all the vitamins in varying proportions. In particular, vitamin B_{12} (the lack of which can lead to pernicious anemia) is universally present, making them a major natural vegetable source for this important substance. They all offer vitamins A, B, and C in quantity (nori is especially packed with these vitamins). Kombu has three times the B vitamins of milk and grains, rivaling the red muscle meats. Although the true form of this vitamin, D_3, resides only in foods of animal origin, such as fish, the extremely high level of ergosterol, one of ten inactive steroids considered as pro-vitamin D, in sea plants (it will be converted to vitamin D in the body) is responsible for the U.S. Government claim that "seaweed oil contains 1,000 times more vitamins A and D than cod liver oil." In the algae we also find highly significant amounts of vitamin E, as well as vitamin K, the antihemorrhagic vitamin. Discovery of the antisterility agent, vitamin S, has led to the beneficial use of kelp on the reproductive organs.

Minerals—the foods of the endocrine glands, which are the vital regulators of our bodily functions, through the production of hormones—are the sea plants' greatest contribution to health. Fully 5 percent of our total body weight consists of minerals. One important mineral, iodine, so scarce in the earth, has been sought as a goiter preventive in sea foods by peoples of all cultures. Among its many vital activities in the thyroid gland, iodine functions as an antiseptic and as such is prophylactic in all bacterial and viral diseases. (Various other algae have antibiotic properties known to be effective against penicillin-resistant bacteria.) Moreover, by maintaining balanced thyroid function and promoting fluid (osmotic) balance in body cells, algae work to counteract obesity. Sea plants also contain the interdependent elements of calcium, phosphorus, and magnesium, which ensure good bones, teeth, nerve transmission, and digestion. In kelp, the potassium-sodium ratio (3:1) more closely approximates that in the body (5:1) than does our most common source of sodium, table salt (1:10,000). These elements also monitor fluid balance. Upsets in the balance of fluids bring about water retention, in addition to disturbances in the lymph and blood pH (the strongest contributors to arthritis). The presence of zinc has led physicians to prescribe kelps in prostate and ovarian disfunctions, including sterility in males. A constituent

---

2. Cholesterol accumulation is a *symptom* of unbalanced nutrition, not the original cause of chronic diseases. Cholesterol is necessary to proper bodily function and as such is found in most tissues. In the case of deficiency the healthy intestinal tract will synthesize it. Fatty acids, lecithin, and several B vitamins act as lipotropic agents preventing the buildup of cholesterol in the liver or arteries.

of insulin, zinc contributes to pancreatic health, assisting in the treatment of diabetes. According to Dr. Walter J. Poires, "Zinc deficiency seems to be one of the causes of atherosclerosis," the loss of elasticity in the inner walls of arteries. Research has also indicated that cancer sufferers excrete high levels of zinc, being unable to retain it in their tissues, where it is instrumental in cellular respiration and serves to transport CO_2 from the tissues into the blood for excretion. Iron, manganese, and chlorine, other essential minerals, are found in sea plants, as are all nutritionally significant elements.

In one of the most exciting recent medical discoveries, alginates from kelp were proven to inhibit the body's absorption of radioactive strontium and cadium up to seven-eighths the radioactive dosage received (see *Nature,* December 1965). Considerable research in this area has been carried out at the Gastrointestinal Research Laboratory of McGill University, Montreal. Indications are that the same alginates also remove strontium 90 which has already been absorbed into the tissues. Furthermore, agar has been shown to work like pectin, bonding with toxic metals such as lead and carrying them out of the body.

Thus, sea vegetables are not only good food; they are also very good medicine. The broad and balanced range of nutriments found in them restores the body to its natural state of health.

Sea Vegetables as Fodder and Fertilizer

In the West (especially in this century, among the scientific community), sea vegetables have been considered almost entirely as food for animals and plants rather than for people, primarily because of their remarkable value as fodder and fertilizer. Animals that feed on sea plants are distinguished by their sleek coats, the sweetness of their milk, and their high resistance to pests and disease. The highly prized wools of Shetland and Orkney are clipped from sea vegetable–nurtured sheep, and kelp added to the ration of hens insures fine-tasting eggs with firm shells. The Overbrook dairy herd in America once held the world's record for milk production—and dried sea plants were an important part of their diet.

Most sea plant fodder is now ingested in the form of commercially prepared meal. Icelanders, the Nordics, and other northern Europeans whose familiarity with sea vegetables goes back at least to Roman times harvest *Laminaria* (kelps), *Ascophyllum* (rockweeds), and *Fucus* (rockweeds) for this purpose. In Norway *Ascophyllum* meal has been shown to increase milk productivity by 6 percent; moreover, since it costs about the same as other commercial preparations, the net result for the dairy farmer is reported to be a 13 percent increase in income.

As a fertilizer, sea plants have rivaled manure in coastal areas of Europe and parts of Asia. In the seventeenth century fertilizer made from brown algae was used so widely that the French government issued decrees regulating their collection; the Brittany and Normandy coast was dubbed the *Ceinture Dorée* ("Golden Belt") for the richness of its sea plant–fertilized crops. Even with the introduction of modern farming methods, Chinese farmers prefer to fertilize their coffee, coconuts, sweet potatoes,

Unloading "driftweed" in Scotland

and ground nuts with species of *Sargassum*. In the nineteenth century the first sea plant harvesting enterprise in the United States, established over 100 years ago, supplied Connecticut tobacco farmers with fertilizer.

"Driftweed" (cast-off fronds or whole plants swept ashore by tide and gale) provides the most accessible source for fertilizer. Commercialization is developing slowly; most users still buy it from hand gatherers or gather it themselves.

Sea plants are superb soil builders. The most soluble substances in soil (including nitrates, chlorides, iodides, bromides, sodium, potassium, calcium, and magnesium) first leach to the sea. Lightning oxidizes atmospheric nitrogen to produce nitric acid, which is carried down by rain and dissolves soil minerals into nitrates. The rain dissolves atmospheric carbon dioxide and carries it down to the soil, where it works to enhance limestone solubility. Today we help the rain and the rivers to wash these important soil vitalizers down to the sea by clearing land and also by dumping garbage into the oceans. However, the marine algae absorb these rich minerals in high concentrations. Sea vegetable carbohydrates decompose more readily in the soil than do the carbohydrates of land plants, and they further the growth of soil bacteria. In sandy soil the colloids, or solid particles that remain in suspension, in sea plants add bulk and help the soil to retain moisture; they soften clay-rich soil and help it to support plant growth. Sea vegetables also contain large quantities of auxins, or plant hormones, which further stimulate plant growth. So powerful are these storehouses of soil nutrients and plant hormones that the Aran Islanders grow potatoes, peas, cabbages, parsnips, carrots, cauliflowers, and other vegetables on stone—by adding sea plants to sand and spreading it on the beds.

Europeans have long appreciated the high resistance of sea vegetable–fertilized plants to insect pests and plant diseases. In America increasing disaffection with toxic pesticides has led to growing interest in sea plants' properties among agriculturalists, horticulturalists, and botanists. Already many new products that include sea vegetables as primary ingredients are being marketed. Sea-based fertilizers are now available for use on house plants as well as crops and greenhouse plants.

Miscellaneous Uses

"Oh, what an endless work have I in hand / to count the sea's abundant progeny!" exclaimed Spenser in *The Faerie Queene*. But even Spenser could never have dreamed of the incredible range of products and by-products now issuing from the sea's plant progeny. So far-reaching are these products that hardly anyone today goes without consuming them daily. The most widespread usage of sea vegetables today, throughout the world, is in the form of extracts. They pop up everywhere, in a range of products from toothpaste to ice cream, salad dressings to house paint.

Species of red algae that yield gelling agents and agar inhabit almost every coastal area. The extract carrageenin, a carbohydrate mucilage obtained from Irish moss (*Chondrus crispus*) and *Gigartina stellata*, gives a softer gel than agar extract; it is used to give body to various dairy products (especially cream cheese, instant puddings, salad dressings, and sauces). Because of its low reactivity with other substances, it is an ideal emulsifier. It is used in cod liver oil and the whole range of beauty aids (creams, lotions, toothpaste, shampoo). Carrageenin is also a water-proofer and fire-retardant; thus, it is ideal for use in varnishes and paints—the casein paint industry alone uses 25,000 pounds per year. Carrageenin stiffens, waterproofs, and gives a finishing gloss to leather; it also serves as a paper and textile thickener. It is found in insecticides, ceramic materials, and rubber. Brewers remove impurities from beer with it. It is even used as a preservative for frozen fish.

Agars are obtained from a variety of red algae. The harder gel of agar has certain advantages; it gels firmly at 37° C and can be reliquified at 85° C; it can be hardened and melted any number of times. Furthermore, it does not "sweat"; it holds moisture and resists bacteria. Agar is used to thicken canned fish and meat, processed cheeses, mayonnaise, and puddings. Bakers and confectioners use it as a thickener and to keep icings from sticking to cellophane. Pharmaceutical companies combine it with lactic acid to produce capsules which counteract intestinal bacteria; in time release capsules it serves both as carrier and laxative agent. Dentists take impressions with agar. Industry includes it in petroleum emulsions and photographic coatings, and uses it as a lubricant when hot-drawing tungsten wire. It even acts as a humidifier for tobacco. Agriculturalists coat plant hormones with it when applying bacteria to the soil through commercial fertilizers. The famous bacteriologist Edward Koch discovered its versatility as a culture medium for analyses of water and milk, or for growing yeast, bacteria, fungi, and molds. Western pathological medicine owes a great deal to agar.

The brown kelps yield the extract algin, which has the largest range of industrial uses. In 1881 the Scotsman E. C. Stanford took out a patent for treating the viscous substance in *Laminaria*, the larger kelps, with sodium carbonate, then with a mineral acid to obtain alginic acid. Although it was not developed commercially until 1934, Stanford's discovery led to a minor revolution in industrial processing. Half the alginic acid produced for the United States is used to make ice cream stable and smooth.

The carbohydrate algin—which aids kelps to withstand pounding surf—generates a "toughness" that is useful in water- and fire-proofing. Alginates are used to strengthen paper and cardboard, to make textiles fire-resistant, and to waterproof cloths (tents and tarps): they are also added to coal briquettes, jellies, cosmetics, plastics, and waterproof paints. Dried *Laminaria* stipes ("sea rods" collected by the seaside-dwelling Irish) are used in the manufacture of glass and soaps. The list of sea plant uses goes on almost endlessly. One such plant even supplies babies in the Hebrides with teething rings.

In the beginning of this century, sea plant burning to extract soda, potash, and iodine for commercial consumption was a large industry in the West; business fell off, however, when

Macrocystis

other, less costly methods of obtaining these inorganic substances were developed. Sea plants are now regaining prominence as new sources of energy are sought. On the West Coast bacteria are applied to *Macrocystis* to convert it to methane gas. Kelps are also being heated to produce petrol-like compounds.

In Japan glues from *Funori,* a species of *Gloiopeltis,* are used in cement, paper coating, and for stiffening textiles. China utilizes 500 tons of *Gloiopeltis furcata* a year for sizing its silks. Cretan sea plant dye which Pliny praised for its colorfast properties is still in use today.

Recently, sea plants have been drafted in the fight against environmental pollution. A red alga (*Lithothamnion calcereum*), formerly used to improve the acidity of the soil, is now introduced into acid drinking water which has begun to corrode pipe metals. The alga neutralizes acids and restores alkaline balance. In large lakes in the United States certain unicellular green algae are actively purifying sewage. Within a few days after innoculation, the algae markedly increase the oxygen content of the water, promoting decomposition of sewage matter. After the water has been circulated for six or seven days, the lakes are cleared of organic matter and pathological bacteria have been reduced to a minimum.

This listing represents only a partial inventory of sea vegetable uses. There seems to be no end to the ways that these ocean plants can be encouraged to benefit humanity and the environment.

Preparing Sea Vegetables

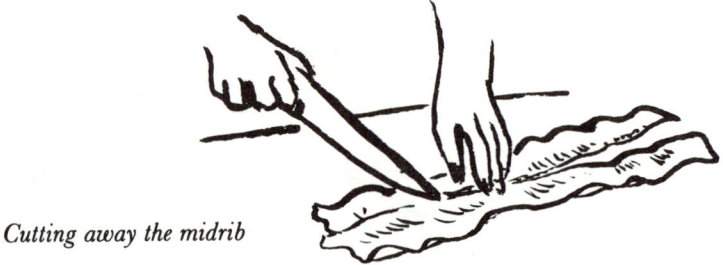
Cutting away the midrib

Sea vegetables are easy to work with and readily incorporated into everyday food preparation.

Where fresh sea plants can be gathered, they need only be brought home, rinsed in fresh water, and used as they are. Fresh, they will keep in the refrigerator as long as any other fresh vegetable. The types of sea vegetables discussed here are available, predried, at many health food stores and Oriental markets. Dried, like herbs, they are best kept in airtight dark glass containers (or in dark places) or in a dry house, hung from the rafters: they will keep in this condition for up to several years. Dried sea vegetables can be easily freshened by merely dropping them in water: they almost regain their original state. The temperature of the water used for freshening is not crucial; it may be room temperature or cooler, but the tougher sea plants freshen more quickly with warmer water (kombu may take up to 15 minutes in cool water, wakame about 4 or 5; more fragile sea vegetables such as nori and dulse, freshen virtually immediately, and often need no freshening at all). Once freshened, sea vegetables will keep in the refrigerator, in soaking water or drained, for about one week.

The bulkier sea vegetables (such as kombu) may simply be removed from the water and, if necessary, patted dry with a towel. Other varieties may be drained through a sieve or large slotted spoon. The soaking liquid may be saved, and, if it cannot be used in the dish being prepared, it can be added to stocks, stews, composts, or fed to pets or plants.

After freshening, some sea plants (such as wakame) may need trimming, as the midrib, or "sinew," of wakame may be a bit tough for delicate dishes. Trimmed away with the tip of a sharp knife, these parts can be saved and added to dishes requiring long simmering, such as soup stocks.

Sea vegetable cooking requires no special tools. A stove, some long-handled cooking spoons, a strainer, pots and pans, are all that is necessary.

Sea vegetables, like all vegetables, are best cooked in flameproof ceramic pots, such as the Japanese *do-nabé,* or fried in woks or heavy iron skillets. Next best is enamel cookware, except for sautéed or stir-fried dishes, which are most efficiently accomplished in a wok. If using a heavy iron skillet, bear in mind that the levels of heat cannot be changed as rapidly as with a wok. When storing cooked sea vegetable dishes (as well as freshened sea vegetables) in the refrigerator, glass or ceramic ware should

Draining arame through a sieved spoon

be used rather than plastic, which may affect the taste of the foods. Pickles and condiments should always be poured into airtight glass containers and refrigerated once they have been opened.

Toasting, Tearing and Cutting: More fragile sea vegetables such as nori and dulse can be made into delicious dishes merely by toasting and serving plain or with a simple sauce or dip. They may be heated on cookie or baking sheets in a moderate (300–350°) oven for several minutes or be passed 1–2 inches over a low flame. After toasting, the fronds can be torn with the fingers or cut with a scissors into thin strips or small pieces, to be mixed together with other vegetables (at the last moment, to preserve their crispness). They may also be strewn on top of almost any dish, from salads to quiches, or used as a seasoning. They should be considered as more than a garnish, for, besides enhancing the flavor of the other foods, they add significantly to a dish's nutritive value and their bulk makes them pleasantly filling.

Powdering: By grinding toasted fronds (with a mortar and pestle, or by mechanical means), one can make a versatile seasoning which can be combined with other spices or used plain, as an alternative/supplement for salt. Certain sea vegetable powders are available commercially (kelp is the most common). Some health food stores also carry powdered sea lettuce and powdered dulse. Japanese food import companies supply nori powder and also nori flakes. The flakes may be sold plain or mixed with other seasonings, such as sesame seeds, poppy seeds, salt, fish flakes, and sometimes (be forewarned) artificial monosodium glutamate. Powdered sea plants can be used as an alternative for salt; they have the added benefit of providing a balanced offering of nutrients. Such powders add a wonderful flavor to dressings, turnovers, croquettes, soups, pot pies, vegetable sauces, and so on. The flakes may be used in addition to, or instead of, bread crumbs or croutons in salads, stuffings, and garnishes of every variety.

Sautéeing: Certain of the sea plants (hijiki and arame in particular), benefit greatly from being sautéed in oil. This process facilitates the digestion of the oil-soluble vitamins while transmuting any overstrong "fishy" taste into a rich sea-vegetable flavor. After the sea vegetable has been briefly sautéed in medium-hot oil, a liquid seasoning mixture (such as soy sauce, honey, and water) may be added and the vegetable left to simmer for 5–10 minutes.

Frying: For tasty snacks, appetizers, or condiments, dried sea vegetables may be quick-fried using a minimum amount of oil. The frying must be done over high heat with the oil hot enough that a seed thrown in will leap up, sizzling. Kombu strips, dulse fronds, wakame strands, and small nori sheets fry very quickly (within a moment for nori and dulse, up to 60 seconds for wakame and kombu): as soon as the color changes, they must be immediately removed and placed onto absorbent paper to drain. The frying produces a change in the flavor as well as in the texture. All become delightfully crunchy, like thin potato chips, with the seasoning built in.

Parboiling: Boiling can leach the vitamins from natural foods, diminishing the taste in the process; it has therefore become less popular among nutrition-conscious cooks. In the recipes presented here, vegetables are rarely dropped into fully boiling liquid; instead, the liquid is brought almost to a boil and the vegetables are poached gently in the near-boiling water, or the liquid is brought to a full boil, the heat is turned off, and the vegetables are added 30 seconds later, once the temperature has dropped. The heat can then be turned on again to simmer. In Japanese cuisine, fresh vegetables (especially greens, but sea vegetables may be used) are dropped into barely boiling water and removed within 30 seconds, or almost instantly, as soon as the color changes to a deeper hue. The vegetables should become tender, but care should be taken to see that they do not lose their crispness altogether. They should then be wrung out to remove as much excess water as possible, chopped by hand coarsely, and served. Several of the sea vegetables, notably dulse, wakame, and nori, can be parboiled and seasoned in this way for tasty salads or side dishes.

Layering: Layered sea vegetables make a delicious accompaniment for steamed or smoked foods and casseroles. Island dwellers the world over place rockweeds in barbecue and smoke pits. In one native American casserole dish, corn is layered alternately with kelp, and tomatoes and chili peppers are placed on top; no more seasoning is necessary. The Hawaiians line their cooking pots with sea plants; New Englanders have always included them in their lobster pots; Alaskans cook them with their mussels. Besides flavoring steamed food or broth with their rich mineral seasoning, sea vegetables—especially the kelps (kombu, wakame, and alaria) and the rockweeds—prevent other ingredients from sticking to the bottom of the cookware. That is why the Japanese place a layer of kombu on the bottom of the pot when cooking rice.

Ingredients

Because of their origins in foreign cuisines, many of the recipes in this book call for ingredients that may be unfamiliar to some people; in other cases natural ingredients are preferable for reasons of taste as well as nutrition.

Miso: Miso is a fermented soybean product with a consistency resembling that of smooth peanut butter. Very high in protein (as are all soybean products) and adaptable to myriad dishes, miso can actually be used interchangeably with peanut butter in most American and many Middle Eastern or African recipes. Miso originated in China: however, it is used far more extensively by the Japanese, who have developed a great number of varieties, with tastes ranging from very salty to quite sweet. The color varies with the taste, from very dark brown to light yellow. In addition to salt and soybeans, grains such as barley and rice may be introduced to give different flavors. (For a thorough description of miso, see *The Book of Miso* by William Shurtleff and Akiko Aoyagi [Autumn Press]). Miso complements sea-vegetables nicely, because the enzymes in the fermented product aid in the digestion of algal starches—and the two taste so good together.

Oils: Cold-pressed natural oils are always preferable to heat-processed oils, since vitamin E, the B vitamins, and other important nutrients are destroyed by overheating. In addition, the body needs the polyunsaturates in natural oils to control its own cholesterol level; overheated oils not only add extra cholesterol to the body but at the same time bind the elements which would allow the body to properly utilize and distribute the cholesterol. Safflower oil is best, as it is highest in polyunsaturates. Sunflower oil is nutritious but has a very strong taste (this can be modified by mixing with safflower oil). Soy oil and untoasted sesame oil are very fine. When using olive oil, many people like the first pressing, or virgin, oil which is highest in chlorophyll, but its powerful taste may again call for blending with a milder oil. Dark, toasted sesame oil, popular in Asia for its appetizing aroma and unique taste, must be used sparingly. When cooking with any oil, one should try to avoid heating the oil any more than absolutely necessary.

Sea Salt: The reasons why powdered seaweed is preferable to sodium chloride have already been indicated. But there are some times when salt is the only thing that will do the job (in pie crusts and sweet desserts, for example), in which cases the natural product, sea salt, should be used. Garlic and cayenne often serve as sufficient seasonings; they are valuable not only for their flavor but for their health-giving properties as well. Inasmuch as all natural seasonings, and for that matter, all wholesome foods, offer more flavors than salt and evoke a healthy flow of saliva and other digestive juices, there is no need to rely on sodium chloride.

Shoyu (natural Japanese soy sauce): An absolute essential in Oriental and much contemporary western cooking, shoyu is high in protein (almost 7 percent) and pleasantly salty; it is a primary component of Worcestershire and other steak sauces, marmite and similar yeast extract meat flavorings and gravy master, all of which take advantage of its ability to enhance the natural flavor of other foods. Traditionally, shoyu is prepared from whole soy beans, using a natural fermentation process. In recent years industrial processes have been developed to speed up the fermentation; so many synthetic additives have been introduced into some common supermarket "soy sauces" that they scarcely resemble the original. Thus, in this book the term natural soy sauce or shoyu is used to refer to the natural product which is being imported from Japan. It is also known in the U.S. by the Japanese name *tamari*. In fact, there has been some confusion over the term, for *tamari* actually refers to a different kind of shoyu that is much heavier and has specialized applications in Japanese cooking. However, the natural food industry is now in the process of rectifying the mistake, and future natural soy sauce will be labelled "shoyu" or "natural shoyu." Shoyu can be used for pickling and is rarely missing from seasoning mixtures and sauces. Shoyu mixed with lemon juice and/or dark sesame oil provides a most satisfactory dressing for sea vegetables, sprouts, and salad greens, while a dash of shoyu and lemon juice added just before turning off the heat brings a wonderfully succulent flavor to sautéed or stir-fried dishes, stews, and steamed foods.

Tahini: The high-protein, high-vitamin, high-mineral mainstay of Near Eastern and North African cuisine, tahini is a thick, usually golden, sauce made from ground sesame seeds. It may be used as is or in combination with lemon juice, shoyu, miso, tofu, vinegar, or honey. It can also be blended into soups or added to wok dishes and sautéed vegetables.

Thickeners: Four thickeners besides agar and Irish moss are used in this book: kudzu, arrowroot, potato starch (in Japanese, *katakuriko*), and brown rice flour. The first three are generally interchangeable, although twice as much potato starch is required to produce a really firm consistency as kudzu or arrowroot. (For a complete treatment of kudzu, see *The Book of Kudzu* by William Shurtleff and Akiko Aoyagi [Autumn Press]). Brown rice flour is useful in soups where the rich grain taste enhances the overall flavor, but it does not thicken as quickly as do the other three. Of all four, potato starch least affects the taste of the liquid to which it is added, but this starch should not be heated over 170°, because it will lose its holding power. Arrowroot will not reheat and should not be fully boiled.

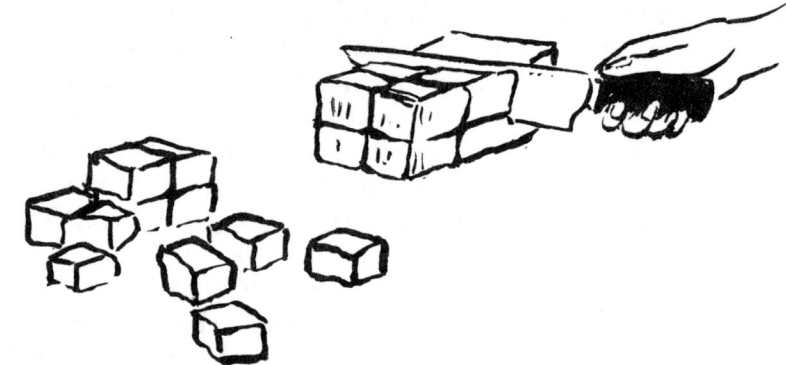

Tofu: Tofu (bean curd), has jumped in popularity recently as a low-cost, high-protein food source and as a meat replacement for partial or complete vegetarians. Cakes of tofu can now be found in supermarkets throughout the country, as well as in Oriental markets and health food stores; many people have begun to prepare tofu at home. When tofu is used in combination with other foods, its taste is rich but not overpowering; it complements sea vegetables in a wide assortment of ways: in dressings, together with sea vegetables in salads, in wok dishes, mashed in tofu-agar pies and aspics. Before sautéeing or frying tofu, it is best to remove as much moisture as possible from the cakes by pressing them: they should be put in a shallow bowl, a plate placed on top of the tofu and a weight on top of the plate. After 15 or 20 minutes, the liquid can be drained from the bowl and the tofu is ready to be patted dry with a towel, cut, and fried. Sizes of tofu cakes vary: Chinese tofu cakes are usually half the size of regular 11- or 12-ounce Japanese portions but more dense. Some pre-packaged cakes available in this country may reach 16 ounces. The recipes in this book use the 8-ounce cakes common on the East Coast, but somewhat larger ones should not affect the balance. (See *The Book of Tofu* by William Shurtleff and Akiko Aoyagi [Autumn Press] for a comprehensive study of tofu.)

Umeboshi (*Japanese salted plums*): Umeboshi offer a surprising range of medicinal benefits. One of the oldest Oriental books on diet relates the seven virtues of umeboshi as: effective antidote; prevents rice fermentation; averts misfortune if eaten regularly; maintains constant taste; promotes bodily endurance; cures headaches; is useful in certain epidemics. They are also a prime preventative and expelling agent for worms and parasites. The citric acid in umeboshi helps the body to absorb calcium from all calcium-rich sea vegetables. Nori is a particularly good companion for umeboshi, because it also aids the digestive processes and it counterbalances the saltiness of the plums. (For a complete medicinal discussion of umeboshi, see *Traditional Herbs for Natural Healing: Umeboshi and Reishi*, by Kosai Matsumoto.) The saltiness of the plums can be diminished by serving with nori in condiments or salads or by combining the plums with another strong sea vegetable such as dulse. The pits may be reserved and dropped into Japanese green tea for a savory, energizing hot drink.

Vinegar: Many washed, fresh sea vegetables or freshened dried sea vegetables can be sprinkled with vinegar and eaten alone, as a salad. Or the vinegared sea vegetables can be mixed with sprouts and other fresh vegetables. The vinegar does double duty, acting as both tenderizer and seasoning. Unrefined, natural products such as rice vinegar, apple cider vinegar, and natural grape vinegars are most healthful.

Sea Vegetables in Meal and Menu

Sea vegetables enhance the cuisines of coastal and island peoples all around the world. The Japanese have explored their gourmet possibilities most thoroughly. The Buddhist stricture against meat-eating, not lifted until the late nineteenth century, led the Japanese to develop the subtleties of vegetable cooking. Many of the dishes that accompany the tea ceremony were borrowed from Buddhist cooking. Tea ceremony hosts vied to create ever finer new dishes. The result was a panoply of refined dishes designed to charm the most discriminating palates.

The Japanese not only offer models in the preparation of many sea vegetable dishes, but can aid our understanding of the importance of aesthetic considerations in planning, preparing, and presenting food. Traditional Japanese food is natural and healthy. Japanese cooks explored the tastes of the natural plants and sea food, rather than masking them with gravies and sauces. Out of certain healthful principles of balance in flavor, texture, and color emerged an aesthetic that places as much emphasis on the decorative appearance of the food as on the taste. Since sea vegetables, grated daikon radish, ginger, and umeboshi plums were known to aid digestion of heavy fats, these came to be included in fried dishes, or accompanied fried foods

as edible decorations. Like parsley sprigs, these garnishes were originally meant to be eaten for nutritional reasons.

In every Japanese meal, textures will be carefully balanced. Colors are varied in the same way. Natural colors indicate the nutrients offered in concentration by different foods, as in the deep iron-red of dulse or the orange carotene (vitamin A) of carrots, and the nutritionally planned menu will include at best two of the color triangle of vegetables—yellow, green, and red-orange.

A harmony of flavors is also sought. Among the five flavors perceived by the tongue (bitter, sour, sweet, hot, and salty), many subtle permutations are possible to the sensitive experimenter. Sea vegetables all taste somewhat salty, yet the iron-mineral taste of dulse is different from the gentle, almost non-sea taste of wakame or the sometimes shrimplike flavor of hijiki or arame.

Woven together in a full composition, the varied themes of flavor, color, and texture do more than please the eye: they encourage the appetite, stimulating the digestive juices to prepare for the nutritious food coming their way. When equal amounts of care and love are invested in the preparation and presentation of food, the result is a work of art, however ephemeral.

II. Cooking with Sea Vegetables

Kombu & The Kelps

The Strength of the Gods

The bursting life force that endows the seaweeds with tremendous powers to sustain and heal is best exemplified by the group of brown seaweeds known as kelps. In over 890 known species—from kombu and the rockweeds to serpent kelp and the great *Macrocystis*—the kelps embody the vibrant potency of nature.[1]

The most fully researched of any seaweed, the *Macrocystis* species are the giants of the plant kingdom, sprawling in lengths up to 1,500 feet. (The tallest land plant, the Douglas fir, towers over everything on land at only 400 feet.) The *Macrocystis* grow as much as 50 feet a year in the cooler waters of the Pacific, where, as Captain Cook noted, they once served sailors as heralds of coastal rocks. These plants have stipes so strong that they are used as rake handles and clubs, though they may be no more than 1/4 inch in diameter. The stipes can grow up to 4 inches a day. Extending lamina (branchings) contain air vessel-floats that keep the fronds bobbing on the surface, allowing them to take in as much light as possible. Fluids are propelled through the stipes at a rate of 20-1/2 inches *an hour*. *Macrocystis* are under consideration as a source of methane gas.

Both *Macrocystis* and the closely related serpent kelps *Nereocystis* are harvested industrially, particularly along the West Coast. There they are processed to procure algin; in some cases they are powdered or pressed into mineral tablets. *Nereocystis* grows in depths of up to 60 feet; its ribbonlike blades reach 16-1/2 feet in length and swarm with spores—6 million per square inch (the water around kelp beds turns milky with these teeming carriers of life). Serpent kelps' huge, bulbous floats may measure 6 or more inches in diameter. One stipe may weigh 24 pounds.

Rockweeds (various species of *Fucus*) are the most widely distributed plants in the world. They turn up everywhere (on rocks and sea shells, exposed pilings, submerged timbers), their labyrinthine jungles providing an environment in which animals and smaller plants are protected against dessication. Natural gelatin guards *Fucus* cells against the perils of the extremes in the

---

1. Wakame and arame are also kelps; however, here they are treated separately, because their taste and culinary roles are so different.

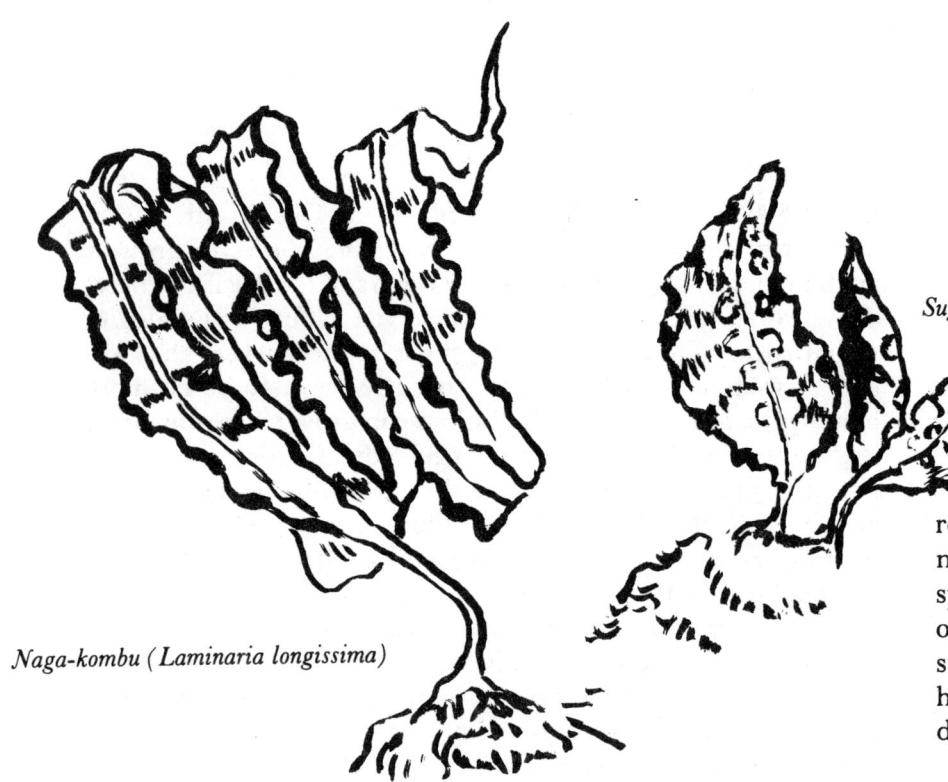

Naga-kombu (Laminaria longissima)

Sugarwrack (Laminaria saccharina)

zone between tide marks: crashing waves at high tide and exposure to sun and wind when the tide is low. When the tide ebbs, they retain water, drying only partially.

Fucus are a source for many kelp tablets, as well as food, fodder, and fertilizer. Roman women beautified their faces with a cosmetic made from *Fucus linneaus*. *Fucus* was in such universal demand that the very word *fucus* was taken from the Greek *phykos*, denoting "algae" as a whole.

The genus *Laminaria* supplies us with kombu, and there must be almost as many species of Laminarians as there are rockweeds. Also known as "oarweeds," Laminarians are perennials; growth begins in the region between the stipe and frond in spring, when the rapid upshoot of young fronds pushes the old ones off the stipe. After much pounding and pulverizing by the surf, the bits of old frond that are left after the plankton and fish have had their fill wash up in the fall storms as "mayweed," or driftweed.

The kelps as a whole have always been regarded with a measure of awe. The Scottish used to warn children to beware of "kelpies," horselike sea creatures with trailing leafy tresses of kelp. Once mounted, these creatures would carry the unwary rider to a watery grave.

The dynamic properties of kelp have long been recognized. Dr. Weston Price, a dentist, traveled to Peru in the 1930s to research the incidence of dental decay in the peoples of the Andes. He himself could not withstand the altitude higher than 12,000 feet. Meeting people who lived at 16,000 feet, he learned that each carried a little bag containing kelp, "to guard the heart." The world-famous Tibetan Sherpas also carry kelp, tak-

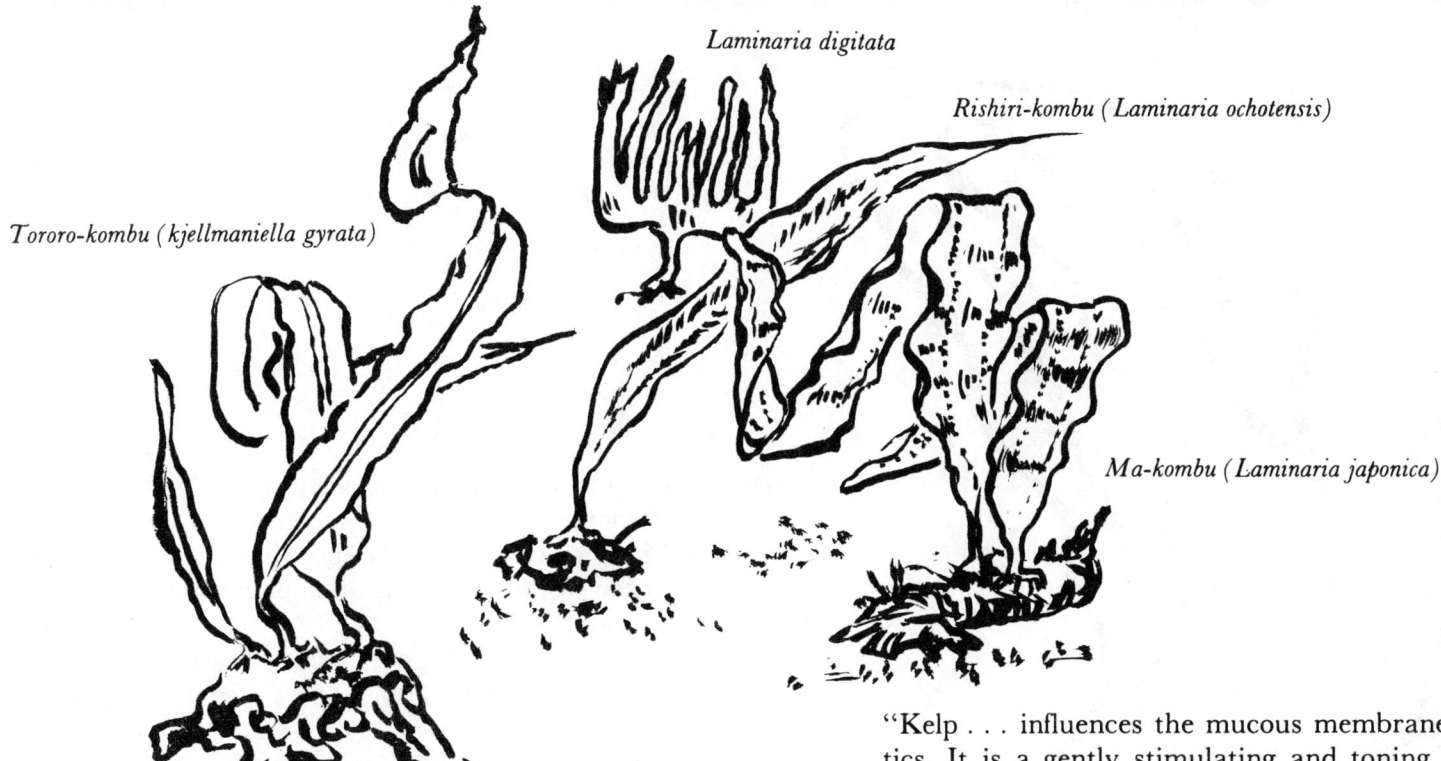

Tororo-kombu (kjellmaniella gyrata)

Laminaria digitata

Rishiri-kombu (Laminaria ochotensis)

Ma-kombu (Laminaria japonica)

ing a pinch at high altitudes to aid their breathing and restore tired leg muscles. They say that seaweeds impart the "strength of the gods."

Dr. D. C. Jarvis, author of *Folk Medicine,* has treated patients for heart pains, rheumatic fever damage, and arthritis with kelp. In 1965 Prof. Tsunematsu Takemoto found that the amino acid laminin found in the Laminarians effectively combats high blood pressure. Since olden times the Japanese have used dishes containing kombu as guardians against high blood pressure, particularly among the elderly. Says Dr. T. J. Lyle,

"Kelp . . . influences the mucous membranes and the lymphatics. It is a gently stimulating and toning alterant . . . stimulating to the absorbents and it especially influences the fatty globules."

This action explains another appealing capacity of kelps: they aid in weight loss, by promoting the balanced absorption and distribution of nutrients. By the same token, they also help to nourish the overly thin. Moreover, fucose and manitol, the natural sugars found in kelp, do not raise the blood sugar level; thus, diabetics may consume it without danger (see *About Sea Foods* by James Hewitt; see also *Prevention,* August 1972). Kelp is also reputed to aid digestion, cleanse the colon (and, in loosening toxins from the colonic lining, thereby relieve related prob-

lems in circulation and nerves), aid kidneys and the urinary tract, relieve anemia, normalize reproductive organs, et cetera.

Dr. J. W. Turrentine of the U.S. Department of Agriculture affirmed fifty years ago, "Of the 14 elements essential to the proper metabolic functions of the human body, 13 are known to be in kelp . . . It should be made available for all people in all lands."

Dried kombu is packaged in strands or in wide, flat sheets. The latter are known as *dashi-kombu,* because they yield a particularly good soup stock (*dashi*). The strands may be cut with a scissors or broken with the hands into appropriate lengths. (With the dashi kombu, scissors are more effective.) To merely soften slightly, wipe with a damp rag. Kombu must be soaked in freshening water for 5 to 10 minutes, however, before it becomes really soft. Once softened, kombu can be cut with a knife and used like any other vegetable. It also lends itself to wrapping, and can even be used in place of pasta (as in lasagna), or grape leaves (as in dolmas).

In its powdered form, available at most health food stores, kelp offers a tangy alternative to table salt, ideal for those on salt-restricted diets, whether by choice or a physician's directive. Kelp powder brings out the flavors of the foods it spices, rather than covering them. When substituting kelp powder for table salt in a standard recipe, use half the amount called for. For liquid preparations, heat the liquid and kelp powder slightly, to help the powder to dissolve. Kelp powder can be kept in a shaker on the table and used like salt.

Kelp Powder Vegetable Bouillon
Serves 5–7

The general rule of thumb is to use half the amount of kelp powder in place of the salt called for in a standard recipe. The powder dissolves completely in liquid when slightly heated.

- 6–8 cups water
- 2–4 cups leftover broccoli, kale, or parsley stems, faded salad greens, soft celery, etc.
- 1 large onion, diced
- 2 medium carrots, diced
- 2 medium potatoes, diced
- 2 bay leaves
- 1 teaspoon cumin seeds
- 1 teaspoon kelp powder
- 2–4 tablespoons natural soy sauce (shoyu)
- 1/2 cup chopped parsley

Combine the first 8 ingredients in a large soup pot and bring to a boil. Reduce heat and simmer, covered, for 3 to 4 hours, or until thick. Before serving, remove the bay leaves and, if desired, the vegetables. Season the bouillon to taste with soy sauce. Distribute the parsley in individual soup bowls, ladle in the hot bouillon, and serve.

Zucchini à l'Italienne
Serves 4

Any vegetable can be tastily pan-fried this way; the zucchini make a wonderfully hearty companion for Italian salad.

- 5–8 tablespoons cornmeal
- 1 tablespoon kelp powder
- 1 tablespoon thyme
- 2 large zucchini, cut into diagonal slices
- 1/2 cup or more olive oil

Combine cornmeal, kelp powder, and thyme in a shallow bowl, mixing well. Press zucchini slices into mixture, coating thoroughly. Fry in oil until golden. Serve hot.

Tangy Almonds
Makes 1 cup

- 1 teaspoon kelp powder
- 2–3 tablespoons natural soy sauce (shoyu)
- 1 cup raw almonds or other nuts

Preheat oven to 300°. Stir the kelp powder into the soy sauce, add almonds, and let stand for 5 minutes, stirring several times to coat almonds evenly. Roast almonds for 8 to 10 minutes, or until lightly browned, turning them from time to time.

Garlicky Pumpkin Seeds
Makes 1 cup

- 1/2 teaspoon kelp powder
- 1/2 teaspoon garlic powder
- 2–3 tablespoons natural soy sauce (shoyu)
- 1 cup pumpkin seeds or other seeds

Prepare as for Tangy Almonds, but oven-roast the seeds for only 4 or 5 minutes.

Kombu Bouillon (*Dashi*)   Serves 4

This flavored soup stock (*dashi* is the Japanese word for "stock") is very basic and can be used in a myriad of ways. In addition to soups, it adds a subtle nuance where liquid is called for in seasoning mixtures, sauces, and vegetable aspics, to name a few.

- 4–5 cups water
- 5 inches dried kombu, broken into 2 or 3 pieces
- 2 tablespoons natural soy sauce (shoyu)
- 1 teaspoon honey (optional)
- 1/4 teaspoon kelp powder or to taste
- 1 scallion, chopped

In a (preferably ceramic) soup pot, heat the water and kombu almost to a boil. Turn down the heat and simmer for about 10 minutes, until the broth is well flavored with the kombu. Add the remaining ingredients and adjust the seasonings to taste. Use your favorite herbs. If allowed to stand, covered, for 20 minutes or more, the flavors will blend more thoroughly. Remove kombu before serving (and save for use in other dishes, such as Kombu Tsukudani or Soy Pickles).

Creamy Avocado & Kombu Soup   Serves 4–6

- 2 tablespoons olive oil
- 1 clove garlic, minced; or substitute the whites of 1/2 leek
- 1/2 teaspoon dill weed
- 3 strands of kombu, each 6 inches long, immersed in 2 cups water for 30 minutes
- 2 cups soy milk
- 2 cups homemade vegetable stock or water with 1/2 teaspoon kelp powder
- 1/2 cup raw cashews (optional), liquified in a blender with 1 cup vegetable stock (Kombu Bouillon)
- 1 small avocado
- 1/2 bunch watercress, liquified in a blender with avocado and 1 cup vegetable stock (Kombu Bouillon)
- 2–3 tablespoons lemon or lime juice, or white wine vinegar
- A sprinkling of freshly ground black or cayenne pepper
- A few sprigs of fresh dill or other herb
- Lemon wedges or rings

Heat the oil in a soup pot, and sauté garlic for 30 seconds (or leek until transparent), stirring in dill weed towards the end. Add kombu and soaking water, cover pot, and cook over medium heat for about 5 minutes, or until seaweed is tender. Remove pot from heat and uncover to cool slightly. Now liquify in a blender or purée in a food mill. Return kombu stock to soup pot and blend in soy milk, cashew/stock mixture, and watercress/avocado purée. Gently reheat. Season with citrus juice or vinegar. Serve sprinkled with pepper and garnished with dill and lemon wedges.

Chinese Cucumber Salad

Serves 3–4

This Chinese salad was meant to be served either hot or cold; refrigerating it for several hours encourages the flavors to permeate the vegetables more fully.

Dressing:

    2 tablespoons natural soy sauce (shoyu)
    2 tablespoons rice vinegar
    1/2 teaspoon honey

14 inches dried kombu, freshened in hot water
3 small or 2 medium cucumbers, cut lengthwise into halves
2 tablespoons dark sesame oil
1 scallion, chopped
1 teaspoon chopped fresh ginger
A sprinkling of kelp powder
A sprinkling of cayenne or shichimi togarashi (Japanese 7 spices; see p. 77)

Prepare dressing and set aside. Pat kombu dry and cut it into 1-inch squares. Using a sharp knife, score cucumbers on the diagonal, as shown, then cut them into bite-size pieces.

  Heat the oil in a wok over high heat. Add cucumbers and stir-fry for 30 to 60 seconds, until evenly coated with oil. (Do not let the cucumbers lose their crispness.) Stir in kombu, scallions, and ginger, then add dressing, mixing well; turn off heat immediately. Sprinkle on kelp powder and cayenne to taste.

  Refrigerate for 3 hours or more before serving. This salad will keep for 2 to 3 days.

  Variations: Use zucchini or squash instead of, or in addition to, the cucumbers. For an especially sweet and delicious salad try the summer squashes known as cymling, pattypan, or scallop squash.

Korean Cucumber Salad

Serves 4

2 or 3 medium cucumbers (or, preferably, 5 or 6 small pickling cucumbers), cut in dime-thin rounds
1 teaspoon sea salt

Dressing:

    2 tablespoons dark sesame oil
    1 teaspoon honey
    1/4 teaspoon kelp powder
    1 tablespoon natural soy sauce (shoyu)
    A pinch of cayenne

1 scallion, minced
12 inches dried kombu, freshened in water, drained, and cut into 1/2-inch squares
1 teaspoon minced fresh ginger
1 clove garlic, minced
2 tablespoons white sesame seeds, toasted in a pan and pounded with a pestle

Layer the cucumber rounds with a sprinkling of salt; allow to stand for 30 minutes, then wash and thoroughly pat dry.

  Prepare the dressing in a large bowl, mixing well. Add the cucumber rounds and all remaining ingredients, and toss together. Refrigerate for 3 to 4 hours or more.

Mediterranean Salad with Kombu

Serves 4–6

In this *à la Grecque* technique, kombu and vegetables are sautéed in oil before dressing is added.

Dressing:

- 1/3 cup olive oil
- 1/4 cup natural red wine vinegar
- Juice of 1 lemon
- Juice of 1/2 orange
- Juice of 1 lime
- 1 large clove garlic, minced or pressed
- 1/4 teaspoon each kelp powder, paprika, and cumin
- A generous 1/4 teaspoon each oregano, dill, mint, and thyme

- 7 feet dried kombu, freshened in water until very soft (simmered briefly, if necessary) and cut into 1-inch-long pieces
- 1/2 cup vegetable oil
- 1/2 pound zucchini, cut into bite-size pieces
- 1 pound eggplant, peeled and cut into large chunks
- 1/2 pound stringbeans, steamed just prior to use
- 1 cup cooked chickpeas
- 6 ounces mushrooms, sliced and briefly steamed just prior to use
- 1/2 cup walnut meats, steamed just prior to use
- Tomato wedges (optional)
- Olives (optional)

Prepare dressing by mixing all ingredients. Combine half of the dressing with kombu pieces in a large glass bowl and set aside to marinate. Heat 2 tablespoons oil in a large pan, and fry zucchini and eggplant in several batches, adding more oil by the tablespoonful as required. Cook quickly to sear and brown surface of vegetables without cooking through. Add fried vegetables to marinating kombu, tossing occasionally. Now add stringbeans, chickpeas, mushrooms, and walnuts. Although it may be served immediately, this salad improves with age. Serve as is, as an appetizer, or arrange over a bed of lettuce and fill out with tomatoes and olives.

Avocado Split

Serves 4

- 3 tablespoons dark sesame oil
- 1 cup mung bean sprouts
- 14 inches dried kombu, freshened in water, drained and chopped
- 2 tablespoons natural soy sauce (shoyu)
- Juice of 1/2 lemon
- 2 ripe avocados, peeled and cut into wedges
- A sprinkling of cayenne (optional)

Heat oil in a heavy (preferably enamel) pan, and stir-fry sprouts and kombu for about 1 minute. Add soy sauce and lemon juice, mixing well, and remove from heat.

Arrange the avocado wedges on a serving platter and top with hot sprouts and kombu. For added pungency, dust with cayenne.

Kombu-wrapped Tofu

Serves 4

This is one of many special Japanese delicacies served in exquisite lacquer boxes on New Year's Day or other holidays.

Seasoning Mixture:

- 1/3 cup sake
- 1/3 cup natural soy sauce (shoyu)
- 1 tablespoon honey

- 24 inches dried kombu (preferably wide "dashi-kombu"), softened in 1-1/2 cups water
- 2–3 tablespoons oil (preferably a combination of safflower and dark sesame)
- 1 cake tofu, pressed, drained well, and cut into 12 pieces
- 12 strips dried gourd, each 5 inches long (or thin strips of kombu), softened in water
- 2 tablespoons sake
- 2 tablespoons natural soy sauce (shoyu)

Combine ingredients in the seasoning mixture and set aside. Drain kombu, reserving the soaking water. Cut kombu into pieces approximately 2 inches long by 5 inches wide.

Heat oil in a wok or enamel pan, and fry tofu until golden. Drain tofu well on absorbent paper, then wrap each piece in kombu and tie with a gourd strip, as shown.

Combine the reserved soaking water and the seasoning mixture in a ceramic pot or deep enamel pan and add kombu-tofu rolls. Bring to a boil, then reduce heat and simmer for 1 to 2 hours, until rolls are fully flavored. Turn the rolls from time to time. Just before removing pan from heat, stir in 2 tablespoons each sake and soy sauce. Allow rolls to remain in the liquid until ready to serve. Delicious hot or at room temperature.

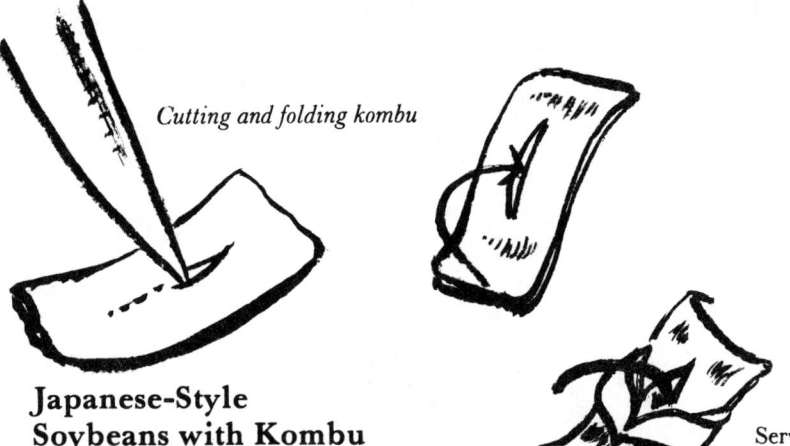

Cutting and folding kombu

Japanese-Style Soybeans with Kombu

Serves 4

3-1/2 cups water
1 teaspoon kelp powder
1 cup dried soybeans, washed
16 inches dried kombu, freshened in water and drained
3–4 tablespoons honey
2 tablespoons natural soy sauce (shoyu)
A sprinkling of savory

In a heavy skillet or enamel pot combine 3 cups water with kelp powder and bring to a boil. Drop in soybeans and return to boil, then turn off heat and let stand for 1 to 2 hours.

Cut the freshened kombu into 12 strips, each approximately 1 inch wide and 2 to 2-1/4 inches long. Make a 1-inch-long cut down the center of each strip and fold the ends through, as shown. Now add kombu to beans and bring to a boil. Pour in remaining 1/2 cup water, reduce heat, and simmer, covered, for 15 minutes. Stir in 2 to 3 tablespoons honey and simmer for 10 minutes more; then add remaining honey and soy sauce, and simmer for 40 minutes. Add a sprinkle of savory and turn off heat. Let stand for 5 minutes before serving.

Black-Eyed Peas with Kombu, Texas-Style

Serves 4–6

2 tablespoons vegetable oil
1 onion, chopped
1 teaspoon chili powder
1/2 teaspoon kelp powder
1 cup chopped tomatoes
30 inches dried kombu, freshened in water, drained, and cut into 1-inch squares
1/2 pound dried black-eyed peas (about 1-1/3 cups), parboiled with little excess water

Heat oil in a heavy skillet or saucepan, and sauté onion with spices, until onion is transparent. Stir in tomatoes, followed by kombu and peas. Cover, reduce heat, and simmer 30 to 45 minutes, or until peas are tender.

Kombu and Sweet Potato
Serves 4

> 6 inches kombu, freshened in water, drained, and cut into thin 1-inch-long strips (about 1/2 cup)
> 4 medium sweet potatoes, cut into 1/2-inch-thick rounds
> 2 cups water
> 4 tablespoons honey
> 4 tablespoons natural soy sauce (shoyu)
> 2 tablespoons black sesame seeds (or poppy seeds)

Place kombu strips, sweet potato rounds, and water in a ceramic cooking pot (or heavy deep enamel pan) and bring to a boil. Reduce heat and simmer, covered, for 10 minutes. Stir in honey, re-cover, and simmer for 10 minutes more. Now add soy sauce, re-cover, and continue to simmer until potatoes are tender (about 3 to 5 minutes). Serve hot, sprinkled with the sesame seeds.

Hot Potatoes and Kombu
Serves 4

> 3 tablespoons natural soy sauce (shoyu)
> 1 teaspoon honey
> 3 tablespoons vegetable oil
> 5 or 6 medium potatoes, cut into bite-size pieces
> 24 inches dried kombu, freshened in water, drained, and cut into 1-inch lengths
> 1/2 cup boiling water

Combine soy sauce and honey, mixing well, and set aside. Heat oil in a heavy skillet or enamel pan. Add potatoes and stir-fry for 2 minutes, or until golden on both sides. Add kombu, then soy sauce–honey mixture, stirring to coat vegetables evenly. Now, add water and bring to a boil. Reduce heat and simmer, covered, until potatoes are soft (about 15 to 20 minutes). Serve hot.

Kombu in Continental Tomato Sauce
Serves 8

When served over millet, pasta, or rice, this side dish becomes a hearty entrée.

> 2–3 tablespoons olive oil
> 1 large onion, chopped (about 1 cupful)
> 1 teaspoon each, paprika and thyme
> 1/2 teaspoon kelp powder
> A pinch of ground allspice, rosemary powder, and cloves
> 1 bay leaf
> 6 ounces fresh mushrooms, sliced
> 1 sweet bell pepper
> 1 large carrot, sliced into thin half-moons
> 1 teaspoon chopped currants
> 3 tablespoons tomato paste, diluted with 1/2 cup water
> 3 cups chopped whole tomatoes (fresh or preserved) or tomato purée
> 14 inches kombu, freshened in 1 cup water, drained (reserving soaking liquid), and cut crosswise into spaghetti-thin strips, or use 1 cup pre-cut kombu strips

Heat oil in a large stove-top casserole and stir in onion, spices, mushrooms, pepper and carrot, in that order; sauté until onion is transparent. Stir in currants, tomato paste, tomatoes, kombu and up to 1 cup of kombu soaking water; simmer for at least 1 hour.

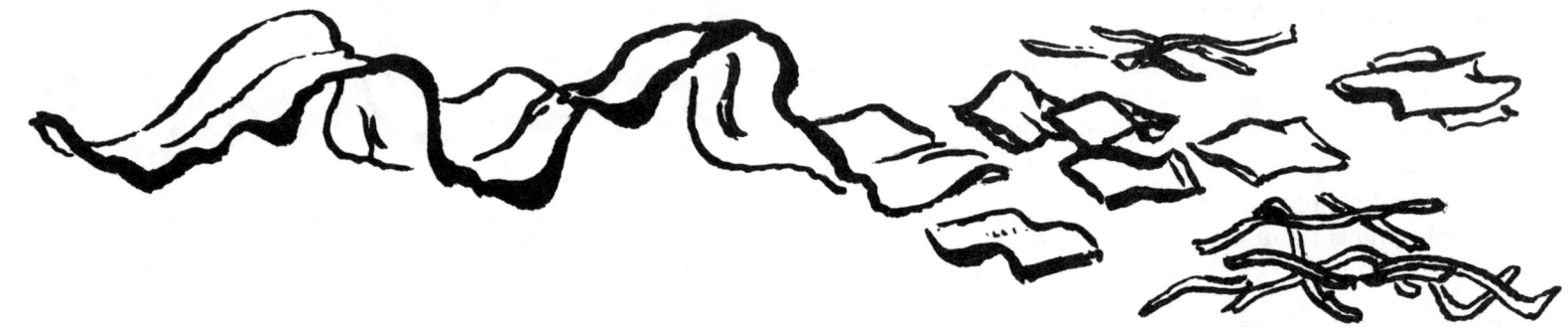

Spicy Kombu Condiment

Makes approximately 1 cup

Use with sprouts as a filling ingredient for sandwiches or a nippy accompaniment for rice.

Seasoning Mixture:

    4 tablespoons natural soy sauce (shoyu)
    2 tablespoons sake
    1/4 cup water
    1 tablespoon honey
    1/2 teaspoon crushed red pepper

2 tablespoons vegetable oil
1 large carrot, slivered
16 snap beans (about 3 ounces), slivered
6 inches dried kombu, freshened in water, drained, and cut into thin 1-inch-long strips (about 1/3 cup)

Prepare seasoning mixture and set aside. Heat oil in a wok or skillet and stir-fry carrots and beans for 30 seconds. Add kombu and stir-fry for 30 seconds more. Pour in seasoning mixture, cover, and simmer until all liquid has been absorbed or evaporated (about 1 hour).

Kombu Tsukudani

Makes 2 to 2-1/2 cups

As with Nori-Mushroom Tsukudani, this rich-tasting condiment can transform even the plainest boiled rice into a flavorsome House Specialty. Or just serve it as a healthy table condiment.

    60 inches dried kombu (preferably thick "dashi-kombu"), freshened in 2 cups water in a shallow bowl
    1/2 cup sake
    2 tablespoons honey
    1/2 cup natural soy sauce (shoyu)

When the kombu is soft, remove it from the soaking water, setting the water aside, and cut it into 1-inch squares. Now combine the squares and the reserved water. Simmer in a medium saucepan, covered, for about 15 minutes. Add the sake and honey, and simmer for 15 minutes more. Add the soy sauce a few tablespoons at a time and continue to simmer over very low heat until almost all the liquid has been absorbed or evaporated, about 6 to 8 hours.

Sunchoke & Kombu Pickles

Makes 1 quart

Sunchoke is a common name for the Jerusalem artichoke, as the plant belongs to the sunflower family. If you can collect fresh kombu, use the stipe or stem portion for particularly crunchy, hollow pickles.

- 4 feet dried kombu, freshened in water for at least 30 minutes and cut into 1-inch-squares
- 1 cup kombu soaking water
- 2 cups apple cider vinegar
- 1 teaspoon yellow mustard seeds
- 1/2 teaspoon black peppercorns
- 1/2 teaspoon each kelp powder, celery seed, coriander seed, cumin seed, and dill seed
- 3 whole cloves
- 6 whole allspice
- 1 whole dried hot, red pepper (optional)
- 4 cloves garlic
- 1 teaspoon honey (optional)
- 10 ounces Jerusalem artichokes, well-scrubbed, trimmed of wilted ends, and thinly sliced (to prevent discoloration, submerge in lemon water until use)
- 1 small red bell pepper, diced

Place kombu, soaking water, vinegar, spices, and honey in a saucepan and bring to a boil; reduce heat and simmer for 15 minutes, then return to a boil. Pack a sterilized quart canning jar with the artichokes and diced pepper. Pour in boiling vinegar-kombu mixture and, using the handle of a wooden spoon, work liquid and kombu down to bottom of jar. Screw cap on tightly and turn jar upside down to ensure sterility of lid. For best flavor, let stand for at least one week. Unopened, the contents will keep over a year. Refrigerate after opening.

Kombu Pickled in Ginger and Soy

Makes approximately 1 cup

Marinade:

- 1/2 cup natural soy sauce (shoyu)
- 3 tablespoons vinegar
- 3 tablespoons sake
- 1/2 teaspoon honey

- 1 to 1-1/2 feet kombu, freshened, drained, and cut into thin strips
- 2 thin, 1-inch-wide rounds of ginger slivered
- 1/3 cup roughly grated carrot

Combine kombu, ginger, and carrot in a sterilized pickling jar, add marinade, and seal; let stand for 3 days. The marinade may be reused indefinitely if the jar is stored in a cool place. Continue to add kombu strips left over from making soup stock.

Cucumbers & Kombu Pickled in Shoyu

Makes approximately 1 cup

12 inches dried kombu, freshened in water, drained, and cut into 2-inch-long, 1/2-inch-wide strips
1/4 cup natural soy sauce (shoyu)
1 medium cucumber, washed, dried thoroughly, and cut into 1/2-inch-thick rounds

Place kombu strips in a sterilized crock or pickle jar; pour in soy sauce, then mix in cucumber rounds. Cover and let stand for at least 24 hours. To serve, drain vegetables. Replenish the ingredients as they are used; the pickling mixture can be used 3 or 4 times.

Garlic and Kombu Pickles

Makes approximately 1 cup

Marinade:

2/3 cup natural soy sauce (shoyu)
1/4 cup sake
1 tablespoon honey
1/4 cup rice vinegar

1/2 pound whole garlic, broken into cloves and peeled (cut larger cloves into halves or thirds)
12 inches or more dried kombu, freshened in water and cut into 1/2-inch squares

In a sterilized jar with a tight-fitting lid, combine garlic and kombu, and add marinade. Seal and let stand for 7 to 10 days. To ensure even absorption of the marinade, shake the jar gently once a day.

If stored in a cool place, these pickles will keep for 2 weeks or more, even after the jar has been opened.

Sweet & Spicy Kombu

Makes approximately 1 quart

Marinade:

3 cups cider vinegar
1/2 teaspoon celery seed
1/4 teaspoon ground mustard
1 teaspoon whole cloves
4 tablespoons honey

4 feet dried kombu, freshened in hot water, drained, and cut into 1-inch squares

Prepare the marinade in a saucepan and bring almost to a boil; reduce heat and simmer for 5 minutes. Place the kombu squares into a sterilized pickling jar and pour in the hot marinade. Seal the jar and let stand for at least one week.

Candied Kombu

Makes approximately 1 cup

1/2 cup honey
1/4 cup water
2 feet dried kombu, freshened in water until very soft, drained, and cut into 1-inch-by-2-inch rectangles

Bring honey and water to a boil, reduce heat, and add kombu. Simmer, uncovered, until almost all the liquid has been absorbed or evaporated (about 1-1/2 to 2 hours). Arrange the kombu rectangles on wax paper to cool, turning them over with chopsticks from time to time to prevent sticking. Serve as a sweet snack.

Deep-Fried Kombu Chips
Makes about 3 cups

These tasty chips are a whiz to make, and it's fun to watch them puff and expand, like flowers opening. A perfect appetizer or snack. Also try deep-frying squares of nori, bits of dulse and sprigs of parsley along with the kombu.

> 66 inches dried kombu
> 2 cups vegetable oil
> A sprinkling of shichimi togarashi (Japanese 7 spices; see p. 77) or cayenne (optional)

With a slightly damp cloth, wipe the kombu gently. Using scissors, cut the kombu into 1-inch squares.

In a deep skillet or wok, heat the oil until a small seed dropped in will sink to the bottom then rise to the top (165° C or 240° F: a little lower than the temperature for tempura). Drop in the kombu and fry until the color changes to a lighter or golden brown, about 30 to 60 seconds. The kombu will expand and little air pockets will rise in it. Remove the kombu from the oil, drain, and sprinkle with pepper, if desired.

Almost any sea vegetable can be made into crispy "chips"

Arame

The *me* of many Japanese seaweed names (like *wakame*) means "young girl," recalling the graceful image of the young seaweed gatherers and the gentle, sinuous plants themselves. Arame, the "rough maiden," appears more robust than the others. *Eisenia bicyclis* is a large Laminarian, thus a kelp, very closely related to wakame and kombu. Arame, however, has stipes that break into radiating bouquets or sprays of fronds which are flat when new, then curl and twist as they grow older.

As a group, the kelps have the most complex cells of any algae. The brownish colorations are all on the surface; the inner cells carry no pigmentation. The way kelps grow is also unique among plants. Growth takes place at the point where the stipe and fronds meet: the outside cells divide and the inside ones stretch, maintaining the same diameter in the stipe at all times.

Arame is a social plant: it grows in convivial association with another brown seaweed, *Ecklonia,* and sometimes, in a temperate climate and open sea environment, with hijiki. Japan is the only country that harvests arame substantially, and only the areas along the main island of Honshu looking out on the Pacific and western Kyushu, the southernmost large island, meet its environmental demands. Inhabiting the variable depths of the open ocean along the coast, arame can grow to lengths of up to 2 meters.

The Japanese have been acquainted with arame for a long time. The name appears in documents kept in the Shoso-in—a treasurehouse of art objects and treatises established by an empress in the seventh century as a memorial to her husband. The tenth-century medical and encyclopedia *Honso Wamyo* notes that arame is useful in treating feminine disorders and mouth afflictions. Both tradition and modern research recommend arame, as well as kombu and hijiki, for high blood pressure. Arame's high concentration of calcium and potassium (among other minerals) undoubtedly contribute to its medicinal properties, which merit further research.

Arame has also been prized as an emergency food. Its ability to maintain its taste, dried, for two or three years, proved a great blessing during a famine in the eighteenth century. *Eisenia* is, moreover, a plant of great untapped potential. A substitute for soy sauce has already been developed from arame and *Ecklonia*. Alginic acid and iodine are obtained from arame, as well as from other kelps.

Of all the dried seaweeds available, dried arame is most dissimilar to its dark yellow-brown, full-fronded appearance in the wild. The wavy 12-by-1-1/2-inch fronds are dried to a charcoal black color, and because the plant is especially tough (its stiff stipe is almost wood-hard), it is chopped into stringlike strips, so that it resembles dried hijiki. With rougher grades of chopped dried arame, it may be helpful to break the longer strands into more manageable lengths before freshening. Chopped dried arame freshens quite quickly in water—5 minutes should suffice—and expands to about double its dried volume. A typical Japanese way of treating arame would be to simmer it (like hijiki) in a soy sauce–sweet sake mixture, or to serve it in miso soup. The rich flavor lends an almost shrimplike sea spicing to tomato dishes, salads, soups, and curries. This sea vegetable is remarkably versatile.

Hot & Sour Arame Soup

Serves 4–6

- 1/2 cup dried arame, freshened in water
- 6–8 (dried) Japanese mushrooms, freshened in water
- 1/4 cup dried "cloud ears" (or "tree ears") mushrooms, freshened in water (optional)
- 2 tablespoons light sesame or peanut oil
- 2 slices fresh ginger root, slivered
- 1/4 teaspoon red pepper flakes or 1/4–1/2 teaspoon cayenne
- 1 stalk celery, sliced
- 5 cups hot water
- 1 cup fresh green peas
- 2 cakes tofu, each cut into 12 pieces
- 2–3 tablespoons brown rice vinegar
- 1 tablespoon natural soy sauce (shoyu)
- 1 tablespoon sake or sweet white wine
- 1 tablespoon arrowroot starch or kudzu powder
- A pinch of Chinese five-spices powder (optional)
- A few drops of dark sesame oil
- A few sprigs of fresh greens (coriander leaves, parsley, or spinach)

In separate bowls immerse arame, dried mushrooms, and cloud ears in water to cover; set aside.

Heat the light sesame oil in a soup pot and sauté ginger, red pepper, and celery for 2 to 3 minutes. Add water, cover, and simmer. Meanwhile, drain arame, mushrooms, and cloud ears, reserving the soaking water; stir arame, mushrooms, and cloud ears into the soup, together with the next six ingredients. Cook gently until peas are just tender. Adjust seasonings, if necessary.

In a small cup, stir arrowroot and spice powder into 1/4 cup reserved soaking water and immediately add to soup. Continue to cook, stirring frequently as soup thickens, about 1 to 2 minutes.

To serve, drizzle a little dark sesame oil into the bottom of each bowl, then ladle in the soup. Garnish with fresh greens.

Broccoli Salad with Arame and Mustard Dressing

Serves 4–6

1/2 cup dried arame, freshened in water and drained
1 pound potatoes (preferably small, new, red potatoes)
1 pound broccoli, cut into bite-size pieces

Mustard dressing:
   1/4 cup safflower oil
   3 tablespoons olive oil
   1 clove garlic, pressed or minced
   1/2 teaspoon mustard powder
   A pinch of rosemary powder
   1 teaspoon marjoram
   1/4 teaspoon kelp powder
   Freshly ground black pepper to taste
   Juice of one lemon
1 head lettuce
1–2 cups alfalfa sprouts
1 small purple onion, sliced into thin rounds

Boil potatoes in water to cover for 10 minutes. Set a steamer filled with the broccoli over potatoes; reduce heat and cook for 5 to 10 minutes, until potatoes are tender. (Do not let broccoli overcook; remove it before it loses its color, and keep warm in the oven.)

While the vegetables are cooking, prepare the dressing by combining all ingredients in the order given; beat lightly. Now drain arame, then mix in 1 tablespoon of the dressing.

To assemble, slice the potatoes and arrange on a bed of lettuce leaves with the broccoli, arame, and sprouts; top with onion rings. Dress and serve immediately.

Sautéed Arame with Soy Sauce and Vinegar

Serves 3 or 4

Here is a simple but delicious way to prepare arame with vinegar, as a salad, side dish, or rice accompaniment.

2 tablespoons vegetable oil
1 cup dried arame, freshened in water and drained
3 tablespoons red wine vinegar or rice vinegar
2 tablespoons soy sauce

Heat the oil in a wok (or enamel pan or skillet), add arame, and sauté for 30 seconds. Stir in soy sauce and turn off heat. Pour in vinegar, mixing well. Serve at room temperature.

Scrambled Tofu and Arame

Serves 4

If the tofu is sufficiently drained, this dish takes on a flavor and texture surprisingly reminiscent of scrambled eggs.

4 tablespoons oil
1 onion, chopped or minced
1/2 cup dried arame, freshened in water and drained
2 medium green peppers, chopped or minced
3 cakes tofu, pressed, drained well, and mashed with a fork
3 tablespoons natural soy sauce (shoyu)
1/2 teaspoon kelp powder

Heat the oil in a deep skillet. Sauté the onion lightly for about 1 minute, until golden. Add the arame and green pepper, then the tofu, stirring well. Cook, uncovered, for 5 minutes. Season with soy sauce and kelp powder, mixing until the soy sauce is evenly distributed. Serve hot.

Brussels Sprouts and Arame Teriyaki

Serves 4

Teriyaki comes from two Japanese verbs, *teru,* meaning "to glisten, to shine" and *yaku,* "to bake, grill, or roast." In this preparation, vegetables, meat, poultry, or fish are braised or sautéed with a sweet sake–soy sauce marinade.

Marinade:

    2/3 cup natural soy sauce (shoyu)
    4 tablespoons honey
    1/2 cup sake
    1/2 cup tomato sauce (or tomato purée with 1/4
        teaspoon pepper)
1 pound Brussels sprouts
2 tablespoons vegetable oil
2/3 cup dried arame, freshened in water and drained
1 tablespoon grated fresh ginger root

Prepare marinade. Using a sharp knife, cut shallow crosses in stems of Brussels sprouts and marinate for 20 minutes.

Heat oil in a heavy skillet and stir-fry sprouts, arame, and ginger for 2 to 3 minutes, then add marinade. Cover pan and simmer for 15 minutes, then uncover and stir-fry to coat vegetables evenly. Serve hot.

Tamale Pie

1 cup cornmeal
2 tablespoons soy flour
1 cup cold water
2 tablespoons vegetable oil
1/2 teaspoon kelp powder
1/4 teaspoon thyme
2 cups boiling water
1 onion, minced
1 clove garlic, minced
1/3 cup dried arame, freshened in water and drained
1/2 teaspoon chili powder
1/2 teaspoon natural soy sauce (shoyu)
1-1/2 cups (2 medium) tomatoes, chopped
1 green pepper, chopped
9 or 10 ripe olives

In a shallow pan, toast cornmeal and allow to cool, then combine with soy flour, cold water, and 1 tablespoon oil to make a thick batter. Stir 1/4 teaspoon kelp powder, thyme, and cornmeal mixture into the boiling water. Simmer over low heat for 1 hour.

Heat 1 tablespoon oil in a separate pan, and brown onion. Add garlic and arame, and sauté for 30 seconds. Season with chili powder, 1/4 teaspoon kelp powder, and soy sauce. Add tomatoes and green pepper, and simmer, covered, until tender (about 4 minutes).

Heat the oven to 350°. Fill half an oiled 8-inch baking dish with the cornmeal mixture; bake for 15 minutes. Now pour in the tomato-arame mixture and layer with remaining cornmeal. Top with olives, and bake for 20 minutes, or until the top is browned to a soft crust.

Arame and Carrots with Hot Peanut Sauce

Serves 4

Hot Peanut Sauce:

    1/3 cup miso
    1/3 cup peanut butter
    1/2 teaspoon shichimi tograshi (Japanese 7 spices; see p. 77)

2–3 tablespoons peanut oil
2 carrots, cut in bite-size pieces
1 onion, coarsely chopped
1 clove garlic, slivered
1 slice fresh ginger root, slivered
1 tablespoon sesame seeds
1 cup dried arame, freshened in cold water and drained
2 cakes tofu, each cut into 10 pieces
1 teaspoon basil
A dash of paprika or cayenne
A dash of kelp powder
1 cup mixed fresh greens of the season, torn or chopped
Water
1/2 pound soba noodles, cooked and drained (reserve 1/3 cup cooking liquid for sauce)
A handful of minced fresh parsley, basil, mint, or coriander (optional)

Prepare the sauce by blending ingredients; set aside to enable the flavors to marry.

    Heat oil in a wok, and stir-fry carrots and onion until carrots are almost tender. Push the vegetables to one side and, adding more oil if necessary, stir-fry garlic, ginger, sesame seeds, and drained arame for 1 minute. Stir entire contents of wok together and top with tofu, then dust with basil, paprika, and kelp. Now layer on greens, splash with water, and cover immediately to trap the steam; lower heat and simmer until greens are wilted and tofu is heated through (3 to 5 minutes).

    Blend 1/3 cup hot soba cooking liquid into the sauce. To serve, mound vegetables on noodles. Top with sauce and, if desired, minced herbs.

Curried Arame

Serves 4

2 tablespoons vegetable oil
1 large onion, cut into long slivers
1 clove garlic, slivered
1 slice fresh ginger root, slivered
1 tablespoon preblended curry powder, or combine the following spices:
    1/2 teaspoon cumin powder
    1/2 teaspoon paprika and/or 1/8 teaspoon cayenne
    1/2 teaspoon basil
    1/4 teaspoon tumeric
    1/4 teaspoon powdered allspice
1 cup dried arame, freshened in just enough cold water to cover

Heat oil in a small wok or skillet, and stir-fry the next four ingredients until onion is partially cooked. Add arame with soaking liquid, and continue to cook until all liquid is absorbed or evaporated.

Arame and Cabbage in Mustard Sauce

Serves 3–4

This tasty dish can be served hot, or allowed to cool and served as a salad or vegetable dish. Cooling and letting the vegetables stand for 30 minutes or more causes the flavors to be more fully absorbed—for those who like their food spicy.

Sauce:

    1/2 tablespoon mustard powder
    3 tablespoons natural soy sauce (shoyu)
    1 tablespoon sake

2 tablespoons dark sesame oil
1 pound cabbage, roughly chopped
2/3 cup dried arame, freshened and drained

Combine sauce ingredients and set aside. Heat oil in a skillet or wok, and sauté cabbage for about 1 minute. Add arame and continue to sauté until cabbage is soft (1 to 2 minutes more). Add sauce and turn off heat. Allow to cool, or serve immediately.

Sweet Cabbage and Arame

Serves 3 or 4

Seasoning Mixture:

    2 teaspoons natural soy sauce (shoyu)
    2 teaspoons honey
    1/3 cup water
    A pinch of kelp powder

1/2 medium head cabbage, shredded
4 cups water with 1/2 teaspoon kelp powder, brought to a boil
2 to 2-1/2 tablespoons safflower oil
2 cakes tofu, cubed
1/3 cup dried arame, freshened in water and drained

Sauce:

    1/4 cup peanut butter
    2 tablespoons honey
    1-1/2 tablespoons natural soy sauce (shoyu)

Prepare seasoning mixture and set aside. Drop cabbage into boiling water and return to boil for 30 seconds; quickly remove pan from heat, and allow cabbage to drain in a colander.

Heat oil in a pan and lightly stir-fry tofu and arame. Now add seasoning mixture and cabbage, and bring to a boil. Reduce heat and simmer, covered, until cabbage is soft (10 to 15 minutes).

While tofu and vegetables are simmering, prepare sauce, thinning to the desired consistency with a little of the seasoning mixture. Stir the sauce into the vegetables just before serving. Serve hot.

Kidney Beans, Snap Beans, and Arame in Tomato Sauce
Serves 4

3 cups water
1/2 pound (1 cup) dried kidney beans, washed
1/8 cup peanut oil
2 cloves garlic, sliced
1 medium onion, chopped
1/2 cup dried arame, freshened in water and drained
1/2 pound snap beans, broken into 1-inch lengths
2/3 cup tomato paste
1/4 teaspoon thyme
1/4 teaspoon oregano
1/2 teaspoon basil
2 bay leaves
1/4 teaspoon kelp powder
1 cup water

Bring water to a boil, add kidney beans, cover, and return to boil. Quickly turn off heat and let stand for 1 hour or more. Then simmer the beans until soft (about 45 minutes).

Heat oil in a heavy skillet, and sauté garlic, onions, and arame for 1 minute. Add snap beans, and sauté for 2 minutes more. Reduce heat to low, stir in kidney beans and remaining ingredients, and simmer, covered, until beans are tender and have become well-flavored (about 30 minutes).

Stir-Fried Arame, Tofu, Carrot and Bean Sprouts with Tahini-Miso Sauce
Serves 3 or 4

Sauce:

2 tablespoons tahini
1 tablespoon miso
1/2 cup hot water
1/2 teaspoon kelp powder
1/4 teaspoon cayenne, or to taste

2 tablespoons vegetable oil
2 cloves garlic, chopped
2/3 cup dried arame, freshened in water and drained
1 cake tofu, cubed
1 carrot, roughly grated
1/2 cup bean sprouts
A handful of parsley, chopped (optional)

Prepare the sauce by combining the first 5 ingredients and set aside.

Heat the oil in a pan or wok and sauté the garlic for a few seconds, then add the arame, tofu and carrot, and stir-fry for about 1 minute. Add bean sprouts and stir-fry for 1 minute more. Turn down heat to medium-low, stir in the sauce, and allow to cook for 2 minutes more. Serve topped with parsley, if desired.

Wakame

The Precious Sea Grass

> *Here it is in Naruto,*
> *a name spoken far and wide,*
> *in eddying tide pools,*
> *that the maidens of the ocean coast*
> *gather the precious "sea grass."*
>
> —The *Manyoshu*

Wakame accounts for 11–15 percent of the total Japanese seaweed harvest (third after nori and kombu). Most is taken from western Hokkaido and northeastern Honshu. The tenderest, tastiest wakame in all Japan is plucked from the furious eddies of Naruto, in the narrow straits separating the southern island of Kyushu from Honshu. Like their close relative kombu, the wakame love the strong currents of turbulent waters, and tender Naruto wakame buds can be nipped two or three times without endangering future generations.

Wakame is a tough plant, but yields readily and deliciously to the loving hand:

> *Willful to the touch*
> *is the wakame of the channel waters*
> *by Tsunoshima;*
> *but to the touch of my own hand*
> *ever gentle and willing is she.*
>
> —The *Manyoshu*

Deeply branching wakame fronds will grow up to 20 inches in water 20–40 feet deep, in sublittoral areas with quick-moving currents. Every spring, when the water temperature goes up, causing wakame to loose vast hordes of spores into the water, the Japanese look forward to enjoying fresh wakame in their morning miso soup.

Alaria is a North American relative of wakame

In North America north of Cape Cod, the winged kelp *Alaria* fills wakame's role. A kombu-like Laminarian, winged kelp consists of a main frond 4–12 inches long with a marked midrib, flanked toward the bottom by multiple "wings" which bear the spores in fall. Like kombu, *Alaria* contains large amounts of vitamin C as well as B_{12}—as much as 60 milligrams per 100 grams. Wakame is a good source of calcium, thiamine, and niacin.

So closely related are wakame and kombu that the dried olive to brownish strands of species of one can sometimes be mistaken for the other, but they unfurl into quite different plants once they are dropped into freshening water. True wakame softens quickly: 2 or 3 minutes is enough, and more than 5 will make the strands too soft. Wakame does not require any special treatment; simply cut with scissors or break into appropriate lengths. *Alaria* must be cooked somewhat longer than wakame.

Both act like leafy land vegetables and make a tasty salad, alone or with a few accompaniments. They add healthful flavor to main dishes and can be used in all kinds of soups.

Clear Soup with Wakame
Serves 4

- 4–5 cups Kombu Bouillon
- 1/4 pound mustard or other greens, washed and cut into 1-inch lengths
- 1/2 teaspoon chopped ginger
- 1 medium scallion, chopped
- 1 cake tofu, cut into thin squares (or 4 pieces *yuba*, or dried bean curd slices)
- 1/2 cup dried wakame cut into 1-inch lengths
- 1 tablespoon sake
- 2 tablespoons natural soy sauce (shoyu)
- 1 tablespoon honey
- A sprinkling of cayenne or shichimi togarashi (Japanese 7 spices; see p. 77)

Bring Kombu Bouillon almost to a boil. Drop in greens, ginger, and scallion, and return almost to a boil. Now add tofu and wakame, reduce heat, and simmer, covered, for 1 minute. Turn off heat and season to taste with remaining ingredients.

Winter Miso Soup with Wakame Serves 4

The Japanese sometimes serve one or another of the many varieties of hearty miso soup two or three times a day. Kombu or kelp are the basis for the stock of the soup the year round, and wakame is a favored addition in any season.

> 4 cups water
> 1 teaspoon kelp powder (or a 2-inch square of dried kombu)
> 2 small potatoes or parsnips, cubed
> 10 strands of dried wakame, cut into 2-inch lengths
> 5 tablespoons miso
> 1-2 tablespoons sake or 1 tablespoon honey (optional)
> A dash of basil
> A handful of parsley, chopped (and/or 1 small scallion, chopped)
> A pinch of cayenne or shichimi togarashi (Japanese 7 spices; see p. 77)

In a pot, combine water, kelp powder (or kombu), and potatoes, and bring almost to a boil: reduce heat and simmer for 5 minutes, or until the potatoes are almost soft, then add wakame. Measure miso into a small bowl and dissolve in 1/2 cup hot broth. Stir dissolved miso into the soup, and season, if desired, with sake or honey. Do not allow the miso to boil. Sprinkle with basil. Chop parsley (and/or scallion) and divide among four bowls. Gently pour in the soup. Serve with cayenne.

Summer Miso Soup with Wakame Serves 4

In summer less miso is needed, and the soup need not be as sweet.

> 4 cups water
> 1 teaspoon kelp powder (or 2-inch square dried kombu)
> 1 medium zucchini or other squash, slivered
> 10 strands of dried wakame, cut into 2-inch lengths
> 4 tablespoons miso
> 1-2 tablespoons sake (optional)
> 1 small scallion, chopped
> A handful of parsley, chopped
> A pinch of cayenne or shichimi togarashi (Japanese 7 spices; see p. 77)

Prepare as for Winter Miso Soup. Serve warm.

A Miso Soup for All Seasons Serves 4

> 4 cups water
> 1 teaspoon kelp powder (or 2-inch square dried kombu)
> 10 strands dried wakame, each 2 inches long
> 2 stalks Swiss chard, chopped fine
> 3 tablespoons miso
> 1 teaspoon honey
> 1 teaspoon basil

In a pot combine kelp powder (or kombu) and water, and bring almost to a boil. Add wakame and Swiss chard. Dissolve miso in a little broth and add, together with honey. Turn off heat. Serve hot, topped with a sprinkling of basil.

Split Pea Minestrone

Serves 4–6

2–3 tablespoons olive oil
2 cloves garlic, minced
1 onion, chopped
2 tablespoons Italian seasoning (or 1 teaspoon each basil, marjoram, oregano, and savory)
1/2 teaspoon dulse or kelp powder
1/2 teaspoon paprika
1/4 teaspoon cayenne
1/4 teaspoon rosemary powder
1-1/2 tablespoons tomato paste
1/2 pound (about 1-1/4 cups) green split peas
2 quarts water
1 medium potato, diced (about 1 cup)
1 large carrot, diced, or cut into rounds and quartered (about 1-1/2 cups)
2 small zucchini (under 1/2 pound), cut into rounds and quartered (about 1-1/2 cups)
8 strips of dried wakame, each 6 inches long, freshened in water (reserve soaking liquid)
1–2 cups precooked white beans or chickpeas
1 cup dried whole wheat elbow macaroni or broken spaghetti
1/4 pound escarole, chopped
1/4 cup minced fresh parsley
Freshly ground black pepper, to taste

Using a large soup kettle, heat as much oil as required to cover bottom with a thin film. Sauté garlic, onion, and spices until onion is transparent. Stir in tomato paste, followed by peas and 1 quart water. Increase heat and bring rapidly to a boil, then simmer, covered, for 30 minutes. Add potato, carrot, and 1 quart water, and continue to simmer for 20 minutes more.

Drain wakame. Pull the soft fronds into small pieces and separate from the tougher midrib, using a knife if necessary. (Set midribs aside for use in other cooking.) Add fronds to soup, along with soaking water, zucchini, beans, macaroni, and escarole. Simmer for 20 to 25 minutes, or until macaroni is tender, then remove from heat. Stir in parsley and dust with pepper before serving.

Tahini and Wakame Soup

Serves 4

2 medium potatoes, cubed
5 cups water
1/3 teaspoon kelp powder
2 tablespoons natural soy sauce (shoyu)
1 teaspoon honey
1/2 cup dried wakame, cut in 1-inch lengths
1/2 cake tofu
5 tablespoons tahini
A dash of shichimi-togarashi (Japanese 7 spices; see p. 77), chili powder, or cayenne

Combine water and potatoes in a large pot and cook until potatoes are tender (about 10 minutes). Add kelp powder, soy sauce, honey, and wakame. Using a fork, break up tofu and add. Dissolve tahini in 1/2 cup broth and stir into soup. Season to taste.

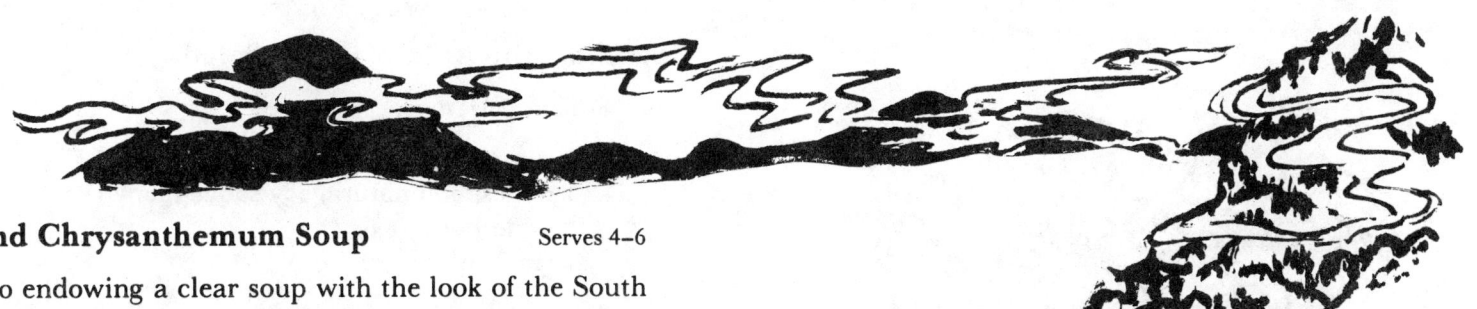

Wakame and Chrysanthemum Soup
Serves 4–6

In addition to endowing a clear soup with the look of the South Seas and the taste of traditional Kyoto cuisine, chrysanthemum blossoms enjoy a long history of traditional physicians' accolades in both China and Japan. Chinese herbal doctors have prescribed them for headache, dizziness, ringing in the ears, eye afflictions, and intestinal pain. The cooling quality of the petals makes them ideal for summer soups. The flowers are thought, like seaweeds, to further longevity, and the two often appear together in festive dishes such as this one.

- 1/4 cup dried chysanthemum flowers
- 1 cup water
- 4 cups Kombu Bouillon (or vegetable broth flavored with sake and soy sauce)
- 1–3 tablespoons honey
- 1/2 cup dried wakame, cut into 1-inch lengths
- 1/4 cup chopped spinach (or other sweet green, such as Brussels sprouts)
- A pinch of grated fresh ginger

Freshen chrysanthemum flowers in water for 10 to 15 minutes, until soft. Pull the petals off the calyxes, discarding the calyxes. In a soup pot, bring broth almost to a boil and add honey. Now add the flower petals, soaking water, and wakame. Continue heating (but do not boil) for 2 minutes, then turn off heat. Add spinach and ginger.

"Foggy Mountain" Soup with Snow Peas
Serves 4

- 2 ounces fresh snow peas
- 1-1/2 cups salted water (use 1 teaspoon sea salt)
- 4 cups water
- 1/2 teaspoon powdered kelp
- 1 tablespoon sake
- 2 tablespoons natural soy sauce (shoyu)
- 1 teaspoon honey
- 1/2 cup dried wakame cut into 1-inch lengths
- 4 tablespoons rice flour dissolved in 1/4 cup water
- A handful of parsley, chopped
- A sprinkling of cayenne

Immerse peas in salted water for 15 minutes, then rinse and drain. Prepare a broth by bringing water and kelp powder almost to a boil; simmer for 3 to 5 minutes. Drop in snow peas and return almost to a boil. Reduce heat to medium and stir in sake, soy sauce, and honey. Add wakame and simmer for 1 minute. Return heat to high and quickly add rice flour–water mixture; cook, stirring constantly, for about 2 or 3 minutes more. Season to taste. Serve topped with parsley sprigs and cayenne.

Wakame Open-Face Sandwich

Serves 1

1 slice of bread (or 1/2 muffin)
1–2 tablespoons miso
1–2 tablespoons tahini or nut butter
1 frond of wakame, freshened and wiped with several drops vinegar or lemon juice
1 slice of tofu
1 slice of tomato
A sprinkling of shichimi togarashi (Japanese 7 spices; see p. 77) or cayenne

To assemble, spread bread with miso and tahini. Wrap wakame around tofu and tomato, and place on bread. Sprinkle on spices and broil briefly before serving.

Bamboo Shoots & Wakame Salad

Serves 3–4

Seasoning Mixture:

1 cup water
1 tablespoon natural soy sauce (shoyu)
1 tablespoon sake
1 teaspoon honey

8 ounces bamboo shoots, drained and cut into half-moon slices about 1/2-inch thick
1 teaspoon rice vinegar
1/2 cup dried wakame, freshened in water, drained, and cut into 1-inch lengths

Dressing:

2 tablespoons rice vinegar
4 tablespoons vegetable oil
1/2 teaspoon kelp powder
1/4 teaspoon cayenne
1 teaspoon basil

A sprinkling of chopped parsley or watercress (optional)

Combine seasoning mixture ingredients in a saucepan over medium heat. Add bamboo and simmer until the shoots take on the rich soy sauce color (about 15 minutes). Remove from heat and pour off any remaining liquid (reserve for later use in soups or stews). Place wakame into a small bowl, sprinkle with vinegar, and set aside.

Prepare dressing, mixing well. Combine wakame with bamboo shoots, and toss with the dressing. Refrigerate for several hours. Serve on a bed of lettuce leaves, garnished, if desired, with chopped parsley or watercress.

Wakame Vinaigrette with Cauliflower and String Beans

Serves 6

Vinaigrette dressing:

    4 tablespoons safflower oil
    2 tablespoons olive oil
    1 tablespoon vinegar
    Juice of 1 lemon
    1 small clove garlic, minced or pressed
    1/2 teaspoon tarragon
    1/4 teaspoon mustard powder
    1/4 teaspoon marjoram
    A dash of freshly ground black pepper

10 strips of dried wakame, each 6 inches long, freshened, drained, and pulled or cut from the midrib into bite-size pieces
1 medium cauliflower, steamed and broken into bite-size clusters
A handful of string beans, lightly steamed and snapped into 1-inch lengths
20 small fresh mushrooms, lightly steamed
1/4 cup walnut or almond halves (optional)
1/4 pound ripe black olives
Freshly ground black pepper, to taste

Prepare dressing by blending ingredients together, and set half aside. Toss remaining half with wakame pieces and set aside to marinate for thirty minutes. Add steamed vegetables and nuts to marinating wakame, adding more dressing as required. Garnish salad with olives and season with freshly ground pepper. Serve immediately, or within 2 to 3 days.

Cucumber, Wakame, and Clear Noodle Salad

Serves 4

Dressing:

    4 tablespoons natural soy sauce (shoyu)
    3 tablespoons rice vinegar
    2 teaspoons honey
    2 teaspoons dark sesame oil

1 teaspoon rice vinegar
1 cup dried wakame cut into 1-inch lengths, freshened in water and drained
3 medium cucumbers, sliced in thin rounds, and enough salted water to cover
4 ounces *harusame*, or clear (cellophane) noodles
3 cups boiling water
1 teaspoon chopped ginger
1 tablespoon sesame seeds

Combine the dressing ingredients, mixing well, and set aside. Sprinkle vinegar over wakame and set aside. Immerse cucumber slices in salted water for 10 minutes; rinse and drain. Drop noodles into boiling water, turn off heat, and let sit for 3 minutes. When noodles are soft, drain in a colander. Cut softened noodles into 4- to 6-inch lengths.

In a salad bowl, combine wakame, cucumber, noodles, ginger, and sesame seeds; toss with the dressing. To enhance flavor, refrigerate for several hours before serving.

Korean Salad

Serves 4

The Koreans believe that the hot water immersion makes the vegetables more permeable to the oil—resulting in a tastier salad.

Dressing:

    3 tablespoons dark sesame oil
    1-1/2 tablespoons natural soy sauce (shoyu)
    1/4 teaspoon cayenne
    1 tablespoon white sesame seeds

1/4 teaspoon kelp powder
3–4 cups water
1 cup soybean sprouts
1 cup mung bean sprouts
2 cups spinach, well washed (cut hard stems away)
1/2 cup dried wakame, cut into 1-inch lengths

Prepare dressing and set aside. Add kelp powder to water in a saucepan and, over high heat, bring almost to a boil. Reduce heat to medium, and keep the liquid just below the boiling point. Drop in each vegetable, one at a time, removing the first with a perforated spoon before adding the next. Remove the mung bean sprouts almost immediately; heat the soy bean sprouts for 30 seconds; heat the spinach for 1 minute, and the wakame for 2 minutes. Arrange the vegetables in a shallow serving bowl, keeping each separate, then pour on the dressing. Still keeping each vegetable separate, turn the vegetables in the dressing so that they are well coated. Refrigerate at least 2 hours.

Variations: Use roughly grated carrot and/or chopped Chinese cabbage or other greens.

Soba Noodles with Wakame, Fried Tofu, and Carrots

Serves 4

Soup Stock:

    8 cups water
    1 piece dried kombu, 6-inch length
    1/3 teaspoon kelp powder
    4–5 tablespoons natural soy sauce (shoyu)
    2–3 tablespoons honey
    2 tablespoons sake

2–3 tablespoons oil (preferably a combination of safflower and dark sesame)
2 cakes tofu, cut lengthwise into halves (as shown)
1/2 cup 2-inch strips dried wakame
1 carrot, slivered
8 Brussels sprouts, halved
4 cups water
8 ounces soba buckwheat noodles
2 large scallions, cut into large diagonal slices as shown
1/2 cup chopped parsley
2 sheets of nori, crisped and cut into thin strips

Prepare stock by heating water, kombu, and kelp powder almost to a boil. Season to taste with soy sauce, honey, and sake, then continue to simmer for 10 minutes or more.

While stock is cooking, heat oil in a heavy skillet. Add tofu and fry over medium heat until golden on both sides. Drain tofu on absorbent paper, then cut into thin strips and set aside.

Add wakame, carrots, and Brussels sprouts to the stock, and simmer, covered, for 10 to 15 minutes, or until the vegetables are tender. In a separate pot, bring water to a boil. Drop in noodles, stirring well. As soon as they become soft (2 or 3 minutes), transfer noodles to stock. (For extra flavor, add a portion of the soba cooking liquid to the soup as well.) Now stir the buckwheat noodles, fried tofu, and scallions into the soup stock, and simmer for 2 minutes more. Serve topped with parsley and nori, accompanied by dipping sauce (two parts miso to one part dark sesame oil, accented with a squeeze of fresh lemon juice).

Soybeans and Wakame Stew Serves 4

    1 cup dried soybeans, soaked overnight in water to cover
    5 dried Japanese mushrooms, freshened in water and drained (cut away hard stems)
    6 tablespoons natural soy sauce (shoyu)
    1-1/2 tablespoons potato starch or brown rice flour
    3 tablespoons vegetable oil
    1/2 cup dried wakame, cut into 2-inch lengths, freshened in water, and drained
    1 carrot, diced
    2-1/2 tablespoons honey
    1/2 teaspoon savory

Turn beans and water into a deep heavy saucepan and simmer, covered, for 15 minutes. Sprinkle mushrooms with 1 tablespoon soy sauce and roll them in potato starch. In a separate pan heat oil and fry mushrooms, together with any leftover potato starch, until mushrooms are golden brown (1 or 2 minutes). Now pour entire contents of the pan into simmering beans, and quickly stir in wakame and carrot. Cook, covered, for 10 minutes. Add remaining 5 tablespoons soy sauce, honey, and savory, and cook until beans are tender and most liquid has been absorbed or evaporated (about 30 to 40 minutes).

Japanese Vegetable Stew *(Noppei)* — Serves 4

Originally one of the delicacies served in the more elaborate tea ceremonies that include dinner, *noppei* brings a touch of the Japanese countryside to a one-pot dish. The somewhat comical word *noppei* is related to *noppera-bo,* denoting a person with a smooth, egg-shaped face and no distinctive features; it refers to the smooth, puddinglike texture of this dish.

Seasoning Mixture 1:

    1 tablespoon natural soy sauce (shoyu)
    1 tablespoon sake
    1/2 tablespoon honey

Seasoning Mixture 2:

    2 cups water
    2 tablespoons potato starch (or brown rice flour)
    1 tablespoon natural soy sauce (shoyu)
    1 tablespoon sake
    1/2 tablespoon honey
    A pinch of kelp powder

4 medium potatoes, diced
2 large carrots, diced
4 ounces (1 cup) snap beans, snapped into halves
1/2 teaspoon kelp powder
8 medium fresh mushrooms, quartered
1/2 cup dried wakame, cut into 1-inch lengths, freshened in water, and drained
1 tablespoon grated fresh ginger

Prepare seasoning mixtures in separate bowls and set aside. Combine potatoes, carrots, and beans in a saucepan, add water to cover, and sprinkle with kelp powder. Simmer, covered, for 10 minutes, or until the vegetables are nearly soft. Add Seasoning Mixture 1 and simmer for 5 minutes more.

Now uncover the pan and stir in Seasoning Mixture 2, together with mushrooms and wakame. Stirring constantly, continue to cook, uncovered, until the broth thickens (2 or 3 minutes). Garnish with grated ginger.

Variations: Add 2 diced turnips and/or 1 chopped burdock root, or 5 to 6 asparagus tips, along with the potatoes, carrots and snap beans. Or add 1/2 cup fresh peas or 1 green pepper, sliced into rounds, together with the mushrooms and wakame.

Carrots and other vegetables can be cut into floral rounds to lend a festive touch

Soft Rice (O-Kayu)
Serves 4 or 5

O-kayu (or *o-jiya*) is the Japanese answer to chicken soup: The Japanese have found that "soft rice" does wonders for intractable digestive tracts. Several salted dried plums (*umeboshi*), pitted, cut into halves and stirred in at the last minute will also give the digestion a helping hand, boost resistance, and add flavor.

- 6 inches dried kombu, broken into thirds
- 5–6 cups water
- 2 cups brown rice
- 1 carrot, slivered
- 1/2 cup dried wakame, broken into 3-inch lengths
- 6 dried Japanese mushrooms, freshened in 2/3 cup water until soft, drained (reserve water), and slivered
- 2 small potatoes, cut into eighths (optional)
- 4 tablespoons natural soy sauce (shoyu)
- A handful of parsley, chopped
- 3 sheets of nori, crisped and cut into 1/2-inch-wide strips
- A sprinkling of cayenne or shichimi togarashi (Japanese 7 spices; see p. 77)

Place kombu pieces at the bottom of a heavy pot, add water, and bring just to a boil. Add rice, cover, and return to a boil. Immediately reduce heat, and simmer for 10 minutes.

Arrange carrots, wakame, mushrooms, and, if used, potatoes in a layer on top of the rice. Add soy sauce and mushroom soaking water. Cook, covered, until vegetables are tender and rice is very soft. Top with parsley. Serve, accompanied by nori strips, along with cayenne and soy sauce.

Vegetable Stew
Serves 4

- 1/2 teaspoon kelp powder
- 1/4 teaspoon oregano
- 1/4 teaspoon thyme
- 3 tablespoons brown rice flour
- 1 large onion, skinned and sliced into rings
- 1/2 cup dried wakame cut into 1-inch lengths, freshened in water and drained well
- 4–6 tablespoons vegetable oil
- 4 small potatoes, roughly chopped
- 1 pound broccoli, roughly chopped
- 1/2 cup water

Combine the first 4 ingredients in a shallow bowl, and dredge the onion and wakame. Heat the oil in an enamel pot (or deep heavy skillet) with lid and sauté onions and wakame for about 2 minutes. Add the potatoes and sauté for 2 minutes more. Now add the broccoli and sauté for 1 minute. Add the water, reduce heat, and simmer, covered, until the vegetables are done, about 20 to 30 minutes.

Scalloped Sunchokes & Wakame
Serves 4–5

Both Jerusalem artichokes and sea vegetables offer diabetics "acceptable" sugars. Thanks to the richness of cashew butter no one will ever miss the potatoes and cream so commonly served in scalloped dishes.

> 12 strips of dried wakame, each 6 inches long, freshened in 2-1/2 cups water, drained (reserve soaking liquid), and torn into 2-inch lengths
> 1 cup soy milk
> 2 tablespoons cashew butter
> 1 tablespoon arrowroot starch
> 1/2 teaspoon kelp powder
> 1/2 teaspoon dill weed
> A dash of cayenne
> A dash of freshly grated nutmeg
> A few drops of vegetable oil
> 1 pound Jerusalem artichokes, scrubbed clean and trimmed
> 6 scallions, cut into 1-inch lengths

Preheat oven to 350°. Set wakame aside. To prepare a sauce, stir into 2 cups of the reserved soaking liquid the next seven ingredients; set aside. Place a small amount of the remaining soaking liquid in the bottom of an oiled 1-1/2-quart casserole with cover. Arrange a layer of artichokes on bottom of casserole, followed by a layer of wakame and one of scallions. Repeat layers until all ingredients are used; top with a final layer of artichokes.

Stir sauce quickly and pour over contents of casserole. Cover and bake for 30 minutes; uncover and bake for 10 minutes more.

Wakame Succotash
Serves 4

In this old favorite, the freshened wakame blends right in with the texture of the lima beans and corn to add a delightful touch of the sea.

> 1 tablespoon vegetable oil
> 1/3 cup dried wakame, freshened in water and drained
> 1 cup lima beans, cooked
> 1 cup corn kernels, cooked
> 1 large green pepper, cut into rings
> 1/4 cup water
> 1/2 teaspoon oregano
> 1/4 teaspoon chili powder

Heat oil in an enamel saucepan or heavy skillet. Add vegetables one at a time, sautéeing each lightly. Now add water, cover, and simmer for 3 minutes. Sprinkle with oregano and chili powder.

Wakame Condiment
Makes approximately 1 cup

> 1 tablespoon vegetable oil
> 1 medium onion, chopped
> 1 cup dried wakame cut into 2- to 4-inch lengths, freshened in water and drained well (cut away hard sinews)
> 1 tablespoon natural soy sauce (shoyu)
> 1 teaspoon sake

In an enamel saucepan or skillet, heat oil and sauté onion until golden, then add wakame and sauté for 30 seconds more. Turn off heat and add soy sauce and sake, mixing well. Refrigerate for several hours before serving.

Wakame and Bamboo Shoots with Miso

Serves 2–3

Seasoning Mixture:

   1 cup water
   A pinch of kelp powder
   2 tablespoons natural soy sauce (shoyu)
   2-1/2 tablespoons honey
   2 tablespoons sake

1 tablespoon vegetable oil
2/3 cup dried wakame cut into 1-inch lengths, freshened in water, and drained
6 ounces bamboo shoots, cut into thin half-moons
1 hot pepper, sliced into rounds
1/2 cup fresh peas
2 tablespoons miso
A sprinkling of white sesame seeds

Prepare seasoning mixture in a small bowl and set aside. Heat oil in a wok or pan and stir-fry wakame, bamboo, pepper, and peas together for 30 seconds; add seasoning mixture and simmer, covered, for 5 minutes. Draw off about 1/3 cup of the cooking liquid and use to dissolve miso. Stir dissolved miso into vegetables. Cook over medium heat for about 20 minutes. Garnish with sesame seeds. Serve hot.

Snow Peas with Creamy Wakame Dressing

Serves 2–3

   1 cake tofu, dropped into boiling water for 30 seconds, cooled, and drained
   4 tablespoons tahini
   1 tablespoon honey
   1 teaspoon kelp powder
   1/2 cup dried wakame, freshened in water, drained (pare tough sinews away), and cut into 1/2-inch lengths
   1/2 cup snow peas, parboiled for 1 minute (or until they turn a deeper green)
   Dash of nutmeg and/or cayenne

In a small bowl prepare wakame dressing by mashing tofu with a fork, then stirring in tahini, honey, kelp powder, and wakame. Arrange snow peas on a bed of lettuce or salad greens, and top with dressing. Season to taste.

Braised Sprouts & Wakame

Serves 3

  1 tablespoon natural soy sauce (shoyu)
  1 teaspoon honey
  1/4 teaspoon kelp powder
  1/2 cup water
  2 tablespoons oil
  1 scallion, chopped
  1/2 cup dried wakame, cut into 1-inch lengths, freshened in water, and drained
  1 pound soybean sprouts

In a small mixing bowl, combine soy sauce, honey, kelp powder, and water, and set aside. Heat oil in an enamel pan or skillet and stir-fry scallion for about 30 seconds. Add wakame and sprouts, and continue to stir-fry until sprouts are evenly heated but still crisp (about 1 minute). Now add the soy sauce–honey mixture. Bring to a boil, then lower the heat, and simmer for 35 minutes. Serve hot.

Lentil Spread

Serves 4

Seasoning Mixture:

  3 tablespoons corn oil
  3 tablespoons natural soy sauce (shoyu)
  4 tablespoons sake
  2 teaspoons honey

1 tablespoon corn oil
12 cloves garlic, skinned and halved
1-1/2 cups dried lentils, soaked for 1-1/2 hours
1 cup dried wakame cut into 1-inch strips, freshened in water, and drained
2 large tomatoes, chopped
1 cup water
8 tortillas

In a small bowl combine seasoning mixture ingredients, mixing well, and set aside. In a heavy enamel pan or skillet, heat 1 tablespoon oil and lightly sauté garlic. Add lentils and sáute for 1 minute. Now add wakame, tomatoes, and water, together with seasoning mixture. Cover and simmer until lentils are tender (about 45 minutes). Serve with tortillas, as a filling.

Nori

Red nori
(*Porphyra tenera*) or laver

A Worldwide Favorite

The rocky coast of Kyushu, Japan's southernmost island, is a painter's dream: a great expanse of azure sea, villages of small low cottages flanked by wind-twisted pines, and mile after mile of cross-hatched bamboo fences—which are not really fences at all, but bamboo ricks: stretchers for the seasonal nori harvest.

Over three hundred years ago, the first bundles of brushwood and bamboo (*hibi*) were sunk into the bottom of Tokyo Bay to make it convenient for nori spores to attach. Today, spores are artificially coaxed along, either by affixing shells containing spores in the proper phase of the reproductive process to rafts, or by adding crushed shell to breeding baths and dipping nets into the baths. Once the nori has been harvested, it is washed in fresh water, chopped fine, and spread on frames in paperlike sheets; the dried sheets are then cut and packaged.

Nori comes in many varieties; however, all species fall into either the genus *Enteromorpha* (green algae), or *Porphyra* (red). Species of *Enteromorpha* are so varied as to seem unrelated: their fronds are all hollow tubes, but some flutter in the water in delicate locks of fine hair, while others are large, flat, and slightly inflated. Some species affix themselves to boat bottoms, much to the disgust of sailors, and become world travelers.

Enteromorpha is a strong plant. Even such a powerful plant as the "oyster thief," *Colpomenia peregrina,* which ravaged oyster beds on the Atlantic coast of France and south Britain early in this century, was driven out by *Enteromorpha*.

The dusky jade-colored *Enteromorpha intestinalis* is known worldwide: it is called *limu eleele* in Hawaiian, *ohashi nori* in Japanese, and link confetti in English. It inhabits the Pacific coast of America and Japan, as well as the American Atlantic coast from Bermuda to Florida. As its collection of names suggests, *Enteromorpha intestinalis* is comprised of chains of single long inflated tubes or sacs, and the fronds are often looped and crumpled like intestines. Mature plants can generate fronds 1 foot long and 1/2 inch wide. The tubes are only one cell thick and do not branch; on warm, sunny days, when photosynthesis is at its height, the hollow tubes puff with gas and the fronds vigorously extend upward. *Enteromorpha intestinalis* prefers to settle between the tide lines in the stiller, brackish water where streams empty into the sea. The Hawaiians would wade into the sand at these estuaries, dip the fine, slippery plants out of the water in buckets, and, back on shore, meticulously clean them with fresh water. Then they would salt the limu, pound or chop it, and eat it fresh as a relish.

Green nori and sea fan

The most popular and widespread type of nori in the world is the purple or greenish or olive-brown *Porphyra: asakusa-nori* in Japanese, laver in English. The flat, roundish sheets grow like semitransparent ruffled fans, undulating in the intertidal zone. At low tide, the 1-inch-to-2-foot wild fronds lie exposed to the sea air—and to hand harvesters. British Islanders use it fresh in salads or boiled like cabbage and other greens.

After the kelps and rockweeds, *Porphyra* is the most widely consumed sea vegetable. The very word *laver* actually denotes "water plants" in Latin and dates British *Porphyra*-eaters back at least to Roman times. To this day, in Wales, the classic accompaniment for mutton is laver or laver "bread" (a kind of green porridge). An Irish equivalent is "sloke," served with potatoes and butter. In South Wales, a dark brown *Porphyra* jelly is heated and served on toast with vinegar or lemon juice. The English enjoy "Black Butter," cooked *Porphyra* coated with oatmeal and fried in butter.

The Japanese have researched nori most thoroughly, for both its taste and its nutrition. *Aonori* (*Enteromorpha*) is a good source for calcium (though not quite as good as hijiki or arame and wakame) and for other important minerals such as potassium, magnesium, phosphorus. It also provides vitamin A (the best asakusa-nori contains as much A as some carrots: 11,000 I.U./100 grams). Nori also contains plenty of C, B_1, and niacin, and is high in protein. Because nori decreases cholesterol and aids digestion, the Japanese often use it in combination with fried foods, as a "light" tangy touch.

Store-bought nori generally comes in packages of ten paperlike sheets, each sheet weighing approximately 3 grams and measuring 17 by 19 centimeters. The sheets can be used as they come, or may be crisped first. Crisp by waving or passing over the flame on a gas range, or by heating in a medium (300°) oven for 2 or 3 minutes. The sheets can then be cut with a scissors into squares of any size or into wide or narrow strips (as for garnishes). The crisped sheets yield easily to the fingers; they can be crumbled over stews, casseroles or salads.

Toasted Nori Sheets

    8 sheets of nori
    4 or 5 tablespoons natural soy sauce (shoyu)
    1 teaspoon lemon juice
    A sprinkling of cayenne or chili powder

To crisp, wave the sheets about 1 inch over a low flame until the sheets change color (be careful not to burn them); or grill the sheets on cookie pans in a moderate oven (300°) for 2 or 3 minutes.

Crisped nori sheets can be cut or torn with the fingers into smaller pieces or thin strips for garnishing salads, wok dishes, or rice. A plate of strips of crisped nori may be kept on the table for use as a garnish. Or crisped bite-sized pieces may be served with a small dish of soy sauce and mustard for dipping.

Crisped nori can easily be crumbled into flakes for dusting fried foods or flavoring batters. They can also be used in place of kelp powder in pan-fried dishes.

Korean-Style Nori

Serves 6–8

2 tablespoons dark sesame oil
2 tablespoons natural soy sauce (shoyu)
A sprinkling of cayenne
6 sheets of nori, cut into halves

Preheat oven to 300°. Combine oil, soy sauce, and pepper, mixing well, and brush onto one side of the nori halves. Arrange nori on a baking sheet and bake for 2 to 3 minutes, or until the nori is crisp and saturated with oil. (Be careful not to burn.) Cut the toasted sheets in six pieces. Stick one toothpick through two pieces of nori and serve as a snack, hors d'oeuvre, or piquant condiment for a meal.

Japanese 7-Spice Seasoning

Makes about 1/4 cup

Nori is a prime taste-maker in one of Japan's most traditional seasonings: *Shichimi-togarashi*. This combination of spices was developed almost three hundred years ago and sold at a tea booth located along the approach to Kiyomizu Temple, one of the most popular temples in Kyoto. The more devoted worshippers practiced sitting half-naked under the cold waterfall at Kiyomizu, so it is no wonder the hot pepper spice gained such eager acceptance! Variations in the ingredients, though not the number, appeared through the years, but nori flakes, red pepper, and hemp seeds remained standard.

1/4 sheet nori, crisped over a flame
1/2 teaspoon poppy seeds
1-1/2 teaspoons sesame seeds, ground
1 teaspoon dried sweet tangerine peel (or orange peel), ground
1/2 teaspoon cayenne
1/2 teaspoon roughly ground peppercorns
1/2 teaspoon mustard seeds, ground

Grind crisped nori into tiny bits between thumb and forefinger. The other ingredients can be ground with a mill or a small mortar and pestle. Mix together well, and put in a shaker, to use as one would cayenne or chili powder.

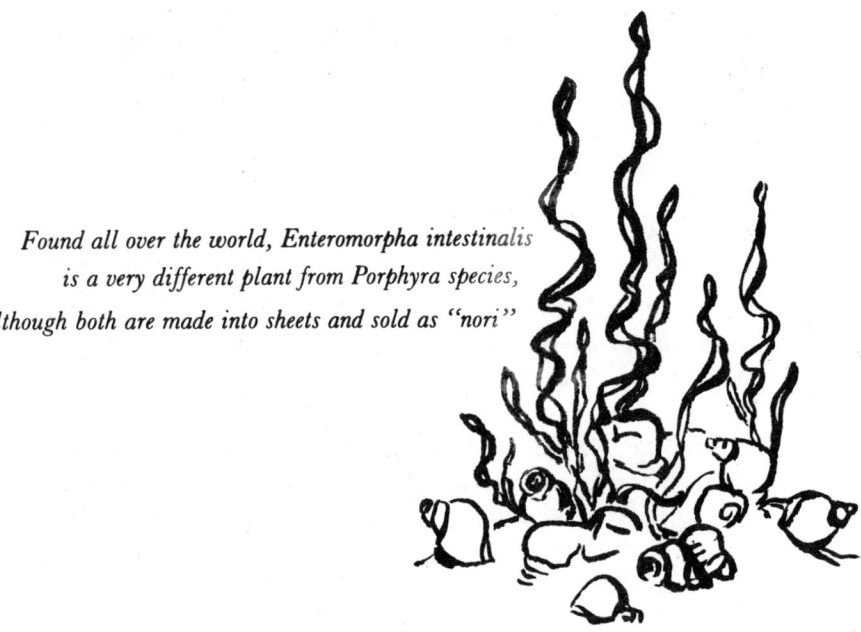

Found all over the world, Enteromorpha intestinalis is a very different plant from Porphyra species, although both are made into sheets and sold as "nori"

Nuts & Nori Hawaiian-Style — Makes about 1 cup

The Hawaiians used to roast candlenuts with nori, but candlenuts are difficult to find at the average supermarket. Almonds work well as a substitute, for snacks or predinner treats.

   1 cup raw almonds
   3 tablespoons vegetable oil
   2 sheets of nori, crisped

Preheat oven to 350°. Mix almonds with oil, coating all the nuts evenly. Arrange nuts on a baking sheet and bake for 5 to 8 minutes. Shred the nori into the roasted nuts, and mix well.

Nori Soup with Umeboshi — Serves 4

A good revitalizer.

   5 or 6 umeboshi plums, pitted and chopped
   4 sheets of nori, toasted and cut into 1-inch squares
   2 tablespoons grated daikon radish
   4 cups boiling water

Distribute the umeboshi, nori, and daikon among four soup bowls. Pour in the water and serve.

African Peanut Soup — Serves 4

   4 cups water
   1/2 teaspoon kelp powder
   1 medium onion, chopped
   1 large leek, chopped
   1 medium yam, chopped
   1/2 cup smooth peanut butter
   3 sheets of nori, shredded
   1/4 teaspoon cayenne

In a heavy 2- to 3-quart saucepan or ceramic soup pot combine the first five ingredients and bring to a boil. Reduce heat and simmer, partially covered, for 40 minutes.

   Purée the soup through a blender or food mill. Return the purée to the pot and return to boil, then reduce heat and allow to simmer. Thin the peanut butter in about 1/2 cup of the broth, then stir into the soup. Add nori and cayenne. Simmer, covered, for 5 minutes more.

Wontons As You Want
Makes approximately 40 wontons

- 10 water chestnuts, chopped fine
- 3 ounces bamboo shoots, chopped fine
- 1 scallion, chopped fine
- 1 stalk celery, chopped fine (or substitute Chinese cabbage or spinach)
- 3 sheets nori, shredded
- 3 tablespoons natural soy sauce (shoyu)
- 1-1/2 tablespoons dark sesame oil
- 2 tablespoons sake
- 3 tablespoons potato starch
- 40 wonton skins

Combine the first five ingredients, mixing well. In a separate bowl, combine soy sauce, sesame oil, sake, and potato starch; now stir the second mixture into the first.

To fill the wontons: place a teaspoonful of the mixture at the center of each wonton skin. Dipping an index finger into a small bowl of water, moisten all four edges of the wonton skin well, then press the edges together. Roll the pressed edges over and pinch hard.

Variations: Wonton Soup: For a delicious wonton soup, prepare "Kombu Bouillon," drop in the wontons, and simmer for 5 minutes.

Fried Wontons: Using a wok or deep skillet, fry the wontons in oil (preferably a mixture of dark sesame and safflower) until golden brown. Serve with a dipping sauce of soy sauce and hot mustard or cayenne.

Steamed Wontons: Steam the wontons and serve with dipping sauce.

Umeboshi, Nori, and Watercress Salad
Serves 4

- 6 umeboshi plums, pitted and mashed or chopped fine
- 2 tablespoons grated daikon radish
- 6–8 drops natural soy sauce (shoyu)
- 1/4 teaspoon basil
- 1 bunch watercress, chopped
- 2 sheets of nori, crisped and cut into matchstick-thin strips

Combine the first four ingredients, mixing well. Toss with watercress and nori strips.

Oriental Cole Slaw   Serves 4

    1 large carrot, shredded
    2 cups shredded Chinese cabbage
    2 cups bean sprouts
    1 tablespoon kelp powder
    1/4 teaspoon cayenne or chili powder
    3 tablespoons dark sesame oil
    2 tablespoons natural soy sauce (shoyu)
    A squeeze of lemon
    A handful of parsley, chopped
    2 sheets of nori, crisped

In a large bowl, toss carrots, cabbage, and bean sprouts; add kelp powder and cayenne, mixing thoroughly. Cover vegetables with a plate that fits down into the bowl and place a weight on top of the plate. Refrigerate for 6 hours or overnight. Drain off liquid and toss vegetables with oil, soy sauce, lemon juice, and parsley. Shred nori on top.

Nori and Cucumber with Grated Radish   Serves 3–4

Because daikon radish and nori help the body to digest fats, this is the perfect salad to accompany fried foods.

    1 large cucumber, quartered lengthwise, then cut into thin slices (as shown)
    1/2 teaspoon sea salt
    Approximately 1 cup water
    1/2 cup grated daikon radish
    3 tablespoons lemon juice
    A pinch of kelp powder
    1 tablespoon honey (optional)
    2 tablespoons natural soy sauce (shoyu)
    2 sheets of nori, crisped

Sprinkle cucumber slices with salt, cover with water, and let stand for 4 to 5 minutes; rinse the slices and pat them dry. Place radish in a mixing bowl, add lemon juice, kelp powder, soy sauce, and honey, then stir in cucumber slices; mix well.

    Shred nori on top of the salad just before serving. Serve accompanied by soy sauce.

Fresh Sprout Salad with Nori
Serves 3–4

Nori adds a subtle sweetness to a basic salad.

Dressing:

    1/3 cup vegetable oil
    1/2 lemon

2 or 3 sheets of nori, freshened, drained, and broken into 1-inch squares
2 cups alfalfa sprouts
1 cup sprouts made from lentils, mung beans, sunflower seeds, or other seeds
1/2 cup grated carrot

Toss nori with sprouts. Sprinkle carrot on top. Serve with dressing.

Variations: This salad invites endless amplification: try cubed avocado, grated beets, shredded cabbage, thinly sliced celery, chopped scallions, watercress, sunflower seeds, nuts. Experiment with dressings.

Fried Tofu, Spinach, and Nori Salad
Serves 4

Dressing:

    1 tablespoon natural soy sauce (shoyu)
    1-1/2 teaspoons dark sesame oil
    1 teaspoon honey
    A sprinkling each of kelp powder and cayenne

2 cakes tofu, pressed and drained well
6–8 tablespoons dark sesame oil
3–4 cups water
1 pound spinach, washed (cut off tough stems)
2 sheets of nori, crisped and cut into thin strips
A sprinkling of lightly toasted white sesame seeds

Blend dressing and set aside. Heat oil in an enamel pan or wok and fry tofu until golden on both sides. Drain on absorbent paper, allow to cool, and then cut into thin strips.

Bring water to a boil over high heat. Drop in the spinach and immediately reduce the heat to medium. Remove spinach as soon as it turns deep green (30 seconds to 1 minute). Pat spinach dry.

Toss spinach with the dressing and tofu. Top with the nori and sesame seeds.

Henry's Favorite
Serves 4

This dish is dedicated to Henry, who said he couldn't eat "seaweed"—and who disposed of an entire bowl of this salad before he found out what he was eating!

Sauce:

    2 tablespoons dark sesame oil
    1 tablespoon natural soy sauce (shoyu)
    1 tablespoon honey
    1 tablespoon white sesame seeds
    A sprinkling of cayenne

8 sheets of nori, crisped and cut into 1-inch strips

Prepare sauce, mixing well. Mix in nori strips, and serve immediately (so as to preserve crispness).

Lentil Salad Ethiopian-Style
Serves 4–5

1/2 pound (1-1/2 cups) dried lentils, washed in cold water
1/2 teaspoon kelp powder in 3–4 cups boiling water

Dressing:

4 tablespoons red wine vinegar
5 tablespoons vegetable oil
1/2 teaspoon kelp powder
A sprinkling of black pepper (preferably freshly ground)

3 sheets of nori, freshened in water, drained and broken into 8 pieces
2 medium scallions, chopped
2 hot peppers, chopped (or 1 Italian pepper, chopped, and 1/2 teaspoon crushed red pepper)

Add lentils to boiling water and cook until tender (20 to 30 minutes). Turn lentils into a colander and rinse with cold water to cool quickly; drain thoroughly and set aside.

Vigorously mix the dressing ingredients. In a salad bowl, combine lentils, nori, scallions, and peppers. Add dressing; stir until well mixed. Marinate at room temperature for 30 minutes or more, stirring occasionally.

Lentil Salad Arabian-Style
Serves 4–5

4 cups water
2 cups lentils, washed
1 large onion
2 whole cloves
2 cloves garlic
2 bay leaves
1 cup chopped onion
1/2 teaspoon cumin powder
1/2 teaspoon powdered cloves
3 sheets of nori, freshened in water, drained, and broken into small pieces
A sprinkling of black pepper
Lettuce leaves
1/2 cup walnut or olive oil
Juice of 2 fresh lemons

Bring water to a boil and add lentils, onion, cloves, garlic, and bay leaves. Reduce heat and simmer, covered, until lentils are tender. Then drain lentils, and remove onion and spices. Combine lentils with chopped onion, cumin, cloves, nori, and black pepper, and arrange on a bed of lettuce leaves. Pour on oil and lemon juice, and toss.

Tofu, Nori, and Nut Butter Salad

Serves 4

Dressing:

    2 tablespoons light-tasting nut butter (cashew, almond, tahini)
    2 tablespoons natural soy sauce (shoyu)
    2 tablespoons rice vinegar
    1 tablespoon dark sesame oil
    1 tablespoon honey

2 cakes tofu, rinsed and drained well, and cut into 12 pieces
1 teaspoon chopped ginger
4 tablespoons chopped parsley
2 sheets of nori, crisped and cut into thin 1-inch-long strips
A sprinkling of kelp powder
A sprinkling of cayenne

Prepare the dressing, mixing the ingredients well. Gently fold the tofu into the dressing, trying to avoid breaking up the tofu pieces. Garnish the salad with ginger, parsley, and nori, and sprinkle on kelp powder and cayenne to taste.

Variation: Sprinkle with crunchy fresh vegetables, such as thin rings of red or green pepper or diced celery.

A Little China, A Little Japan in a Wok

Serves 4

2 tablespoons kudzu powder or arrowroot or potato starch
2 tablespoons natural soy sauce (shoyu)
1 cup water
8 dried Japanese mushrooms, freshened until soft in 1/4 cup water, drained (reserving the water), and slivered
3 tablespoons oil
2 cloves garlic, chopped
2 cakes tofu, each pressed, drained, and cut into 10 pieces
1 pound asparagus, roughly chopped
1 pound or 1 medium head cauliflower, broken into bite-size pieces
2 tablespoons tahini
1/2 cup raw cashews, chopped or roughly ground
4 sheets of nori, crisped and cut into thin strips

Combine kudzu powder, soy sauce, and water. Stir in mushroom soaking water, and set the mixture aside.

    Heat oil in a wok and stir-fry garlic. Add tofu pieces, browning first on one side, then the other. Add asparagus, mushrooms, and cauliflower, and stir-fry until the asparagus color deepens (2 or 3 minutes). Now add kudzu–liquid mixture, stirring occasionally until it begins to thicken; stir in tahini and continue to heat for 1 minute more. Turn up heat for a moment, then turn heat off. Serve topped with cashews and nori strips.

Nori-wrapped Fried Tofu Serves 4

Dipping Sauce:

    4 tablespoons grated daikon radish
    4 tablespoons natural soy sauce (shoyu)
    A squeeze of lemon

or:

    1/4 cup natural soy sauce (shoyu)
    1/2 teaspoon mustard powder

1/4 cup vegetable oil (preferably a combination of safflower and dark sesame)
2–3 cloves garlic, sliced
4 cakes tofu, each pressed, drained well, and cut into 8–10 pieces
10 sheets of nori, toasted and cut into quarters

Prepare dipping sauce and set aside. Heat oil over medium heat in a wok or deep skillet. Add garlic and let brown. Add tofu and fry until golden on both sides. Drain well on absorbent paper and allow to cool to room temperature. Wrap one piece of tofu in each strip of nori.

Nori-Tofu with Rice Serves 4

8 tablespoons oil (preferably a mixture of safflower and dark sesame)
2 cakes tofu, each pressed, drained well, and cut into 8–10 pieces
1-1/2 cups kombu stock (Kombu Bouillon) or 1/4 teaspoon kelp powder dissolved in 1-1/2 cups water
1 tablespoon natural soy sauce (shoyu)
2 scallions, chopped
5 cups hot cooked rice
2 sheets of nori, toasted and shredded or cut into thin strips
2 tablespoons lightly toasted white sesame seeds
A sprinkling of cayenne or shichimi-togarashi (Japanese 7 spices; see p. 77)

Heat oil in an enamel pan or wok. Fry tofu pieces until golden on both sides, and drain thoroughly on absorbent paper, then cut into thin strips. Combine kombu stock and soy sauce, and bring almost to a boil. Drop in the tofu and heat for 30 seconds; then drop in the scallions and turn off heat.

Turn servings of hot rice into heated individual bowls. Pour on the tofu–scallion sauce and top with nori strips, sesame seeds, and a sprinkling of cayenne.

Nori Rolls (*Nori-Maki*) — Serves 4–6

- 1 cup brown rice, washed and soaked in 2-1/2 cups water for at least 30 minutes
- 4 inches dried kombu
- 2 tablespoons rice vinegar
- 1/2 teaspoon kelp powder
- 4 sheets of nori

Filling:
- 1/4 cup grated cucumber and/or carrot
- A dash of soy sauce
- A sprinkling of sesame seeds

Place kombu at the bottom of the rice pot, and cook rice as usual. When done, turn cooked rice into a bowl and allow to cool for several minutes, or until it is no longer steaming. Mix in vinegar and kelp powder.

Place a sheet of nori on a small bamboo mat (or use a heavy cloth napkin). Spread one-quarter of the rice across most of the nori sheet, leaving a 2-inch edge uncovered on all sides. Arrange the filling ingredients in a line across the middle of the rice. Roll the nori in the mat and press gently to make firm. Leave the roll in the mat for 1 minute, then cut into 1- to 1-1/2-inch-wide rounds. Repeat the process with the remaining nori sheets until all ingredients are used.

Variation: For filling use Nori-Mushroom Tsukudani, Kombu Tsukudani, or Spicy Sliced Kombu.

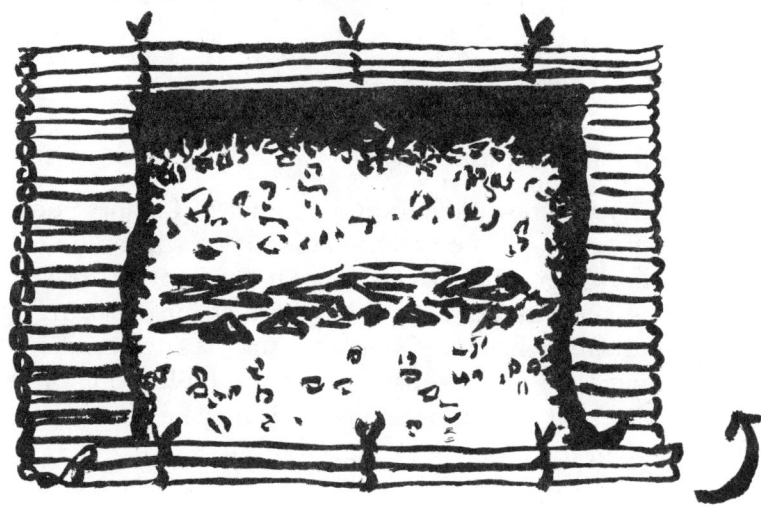

Nori-Mushroom Tsukudani — Makes approximately 1 cup

- 12 sheets of nori, cut into 1-inch squares and freshened in water; reserve soaking water
- 8 dried Japanese mushrooms, freshened in water and slivered; reserve soaking water

Sauce:
- Water
- 4 tablespoons sake
- 3 tablespoons honey
- 6 tablespoons natural soy sauce (shoyu)
- 1 teaspoon fresh ginger, slivered

Add enough fresh water to the leftover nori and mushroom soaking water to make 3/4 cup. Combine the sake, honey, soy sauce and ginger in a saucepan with a tight-fitting lid; stir in the water. Add the nori and mushrooms and simmer, covered, until almost all liquid is absorbed or evaporated, 2 to 3 hours. Pour over freshly cooked rice, or serve as a side dish. Refrigerated in a well sealed jar, Tsukudani will keep for several days.

One-Pot Sushi (*Chirashi-Zushi*) — Serves 4–5

Sushi is one of the most famous and popular of Japanese delicacies, and yet many people do not know that sushi can be made without fish, raw or otherwise—or that it need not be pressed or rolled. It is the vinegared rice that makes sushi sushi: *su*=vinegar, and *shi*=delicious. *Chirashi* means "scattered, like falling flower petals"; in this dish dainty vegetables, bits of fish, ginger pickles, and sesame or poppy seeds may be scattered delicately on top of the vinegared rice.

- 2 cups brown rice
- 4 to 4-1/2 cups water
- 4 inches dried kombu
- 5 tablespoons rice vinegar
- 2 tablespoons honey
- 1 teaspoon kelp powder
- 2 sheets nori, crisped over a flame and cut in thin strips
- A handful of parsley, chopped
- 3 tablespoons lightly toasted white sesame seeds

Wash rice, cover with water, and let stand for at least 30 minutes before cooking. Place kombu at the bottom of the cooking pot (preferably ceramic; otherwise use enamel or heavy iron) and pour in rice and water. Cook rice as usual.

When the rice is ready, turn it into a wooden or ceramic bowl, and add vinegar, honey and kelp powder, mixing well. (If desired, chop the cooked kombu and stir it into the rice too.) Allow the rice to cool to room temperature. Serve topped with nori, parsley, and sesame seeds.

Tsimmas with Nori — Serves 4–5

Tsimmas—or *tzimmas* or *tsimmes,* from the German *zum* ("to the") and *essen* ("eating")—is an Eastern European Jewish dish of cooked vegetables and fruits. Because tsimmas cooks slowly and takes rather a long time, the word came to mean "a big fuss or bother" or "troubles and difficulties"—as in, "Don't make such a big tsimmas out of it!" The dish itself is actually little trouble—and open to numerous variations.

- 4 tablespoons corn oil
- 3 medium carrots, roughly chopped
- 2 medium yams or sweet potatoes, roughly chopped
- 1 large onion, chopped
- 1–1-1/2 cups water
- 6–8 prunes, pitted and halved
- 2–3 bay leaves
- 4 cloves
- 1/4 teaspoon kelp powder
- 1 cup dark bread crumbs
- A squeeze of lemon juice

Topping:

- 2 tablespoons safflower oil
- 1 cup pumpkin seeds, almonds, cashews, sunflower seeds, or any other combination of seeds and nuts
- 2 sheets of nori, crisped

Heat corn oil in a deep heavy skillet or enamel pot and sauté carrots, yams, and onion until onion turns golden. Add water, prunes, and spices, and simmer, covered, until vegetables are very soft (30 to 40 minutes). Mix in bread crumbs and add a squeeze of lemon juice.

In a separate pan, heat safflower oil and sauté nuts and seeds; break the nori into the nuts, mixing well. Sprinkle the nuts-and-nori mixture on top of the vegetables. Serve hot.

Variations: Other dried fruits may be used, such as figs, apricots, and/or raisins.

Greens Topped with Sesame Seeds and Nori

Serves 4

Almost any green benefits from gentle treatment in this manner. Try Chinese rape or green cabbage, if available. Mustard greens or spinach are equally agreeable.

    4-5 cups water
    1-1/2 to 2 pounds greens, washed and chopped in half
    6-8 tablespoons natural soy sauce (shoyu)
    2 sheets of nori, crisped and cut into thin strips
    4-6 tablespoons lightly toasted white sesame seeds

Bring water to a rolling boil, reduce heat to medium, and drop in the stem halves of the greens. Return to boil for 30 seconds, then add the top half of the leaves. Remove leaves as soon as their color turns. (Do not allow leaves to wilt.) Squeeze leaves between palms to remove excess liquid, then distribute in individual bowls. Season with soy sauce and top with nori and sesame seeds. Best served cold.

Variations: For a subtler taste, kombu bouillon may be added to the soy sauce (use approximately 2 tablespoons per serving). Try substituting poppy seeds for the sesame.

Nori on the Side

Serves 3-4

    10 sheets of nori, broken into 1-inch-long pieces
    2 cups water
    1/4 teaspoon kelp powder
    3 tablespoons sake
    2/3 tablespoon honey
    4 tablespoons natural soy sauce (shoyu)
    A pinch of savory or basil

In a measuring cup freshen nori in water, then remove to another bowl, draining well. Add enough water to the measuring cup to again make 2 cups, and pour into a small enamel or ceramic pot. Add the next four ingredients and bring almost to a boil. Drop in the nori pieces and reduce heat, simmering for about 1 minute.

Divide nori and sauce among individual serving bowls, and top with savory or basil. Delicious hot or cold.

Zen Tempura

Serves 4

Dipping Sauce:

    1/4 cup natural soy sauce (shoyu)
    2 tablespoons lemon juice
    1-1/2 cups water
    1/4–1/2 cup grated daikon radish
    1 tablespoon honey
    1/4 cup sake
    1/4 teaspoon kelp powder

1 cup flour (preferably whole wheat pastry)
1 cup water
2 tablespoons mustard powder
1 teaspoon kelp powder
2–3 cups vegetable oil
3 sheets of nori, each cut into eighths
10 strips dried wakame, each 4 inches long, freshened in water, drained, and knotted as shown
1 large sweet potato, cut into rounds
2 medium green peppers, cut into large slices
1 or 2 onions, cut into rounds

Blend dipping sauce and set aside. Combine flour, water, mustard powder, and kelp powder to make a thin batter. Heat oil in a wok or deep skillet until a drop of batter will leap, sizzling, to the surface. Dip nori into batter, coating thoroughly, then drop into oil and deep-fry until golden; drain on absorbent paper. Repeat with remaining vegetables, skimming the oil surface of debris with a strainer, until all are used. Serve tempura as fresh from the frying pan as possible.

Variation: Many other vegetables lend themselves well to deep frying: asparagus tips; bite-size pieces of cauliflower or broccoli; tofu, well drained and cut into small pieces; zucchini or other squash, cut into diagonal slices; slivered carrots, deep-fried by the handful; snap beans; snow peas.

Hot Sauce with Umeboshi and Nori

Makes approximately 1 cup

This tasty condiment will serve as a spicy vegetable dip or as an accompaniment for rice.

    4-1/2 tablespoons peanut oil
    4–5 small scallions, minced
    10–12 umeboshi plums, pitted and finely chopped
    3 tablespoons water
    1/2 teaspoon cayenne (or to taste)
    3 sheets of nori, crisped

Heat oil in a small pan; sauté scallions very lightly, then immediately remove pan from heat. Transfer contents of pan to a bowl, and add plums, water, and cayenne. Crumble in nori, mixing well. If stored in a cool place, sauce will keep for at least 1 week.

Nori-Miso "Pickle" *Makes approximately 2 cups*

3 sheets of nori, cut into 1-inch squares
2 cakes tofu, each sliced into eighths
1-1/2 cups miso, approximately

Using a wide ceramic or glass covered refrigerator dish, layer nori, miso, tofu, and miso, alternately; repeat until all ingredients are used. (The miso need only be thick enough to cover.) Refrigerate for 7 days; the tofu will take on the consistency, and something of the taste, of a delicious, thick cream cheese.

To serve, scrape the miso from the nori and tofu. Use the miso to make one more batch of pickles, or in other dishes.

Umeboshi and Nori Condiment *Makes approximately 1 cup*

7–8 sheets of nori
1/2 cup water
2 umeboshi plums, pitted and mashed or chopped fine
7–8 drops natural soy sauce (shoyu)
1 teaspoon white sesame seeds

Tear nori into small (dime-sized) pieces. Freshen in water and drain, reserving the water for use in other dishes. Combine all ingredients, mixing well.

Tapioca Pudding with Nori *Serves 4–6*

1 pound apples, peeled, if desired, and cored
1 teaspoon light vegetable oil
1 tablespoon honey
A sprinkling of ground coriander and cinnamon
A dash of powdered cloves
2-1/2 cups boiling water
1/3 cup quick-cooking tapioca
3 tablespoons honey
3 sheets of nori, torn into small pieces
1 teaspoon crushed, dried mint leaves (or 1/4 teaspoon mint extract)
2 tablespoons orange flower water (or 1/4 teaspoon orange flavoring extract)
Cinnamon, to taste

Preheat oven to 350°. Coat a small, covered casserole with the oil. Slice apples directly into the casserole. Add next four ingredients and bake for 15 minutes.

Bring water to a boil in a heavy saucepan or double boiler. Stir in tapioca and whisk thoroughly. Cook over medium-high heat, stirring constantly, for 3 minutes. Add next three ingredients and whisk smooth. Remove pan from heat and stir in orange extract.

Now pour tapioca mixture into casserole, covering apples. Bake until apples are tender (about 20 minutes). Serve hot or cold, dusted with cinnamon.

Variation: This dish takes on a different, but equally delicious, flavor when dulse (1 cup dried dulse fronds) is used instead of nori.

Nori Crêpes

Makes approximately 20 6-inch crêpes

Crêpe batter:

2 to 2-1/2 cups water
1 cup whole wheat pastry flour
2 sheets of nori, crisped
3 tablespoons vegetable oil
A dash of freshly grated nutmeg

Sweet Miso-Tahini Filling:

6 tablespoons (rice) miso
6 tablespoons tahini or cashew butter
3 tablespoons apple cider
3 tablespoons honey
1 teaspoon freshly grated orange rind
A pinch of ground allspice
3 tablespoons finely crushed walnuts (optional)

To prepare batter, beat water into flour and nutmeg in a mixing bowl, using just enough water to form a very thin batter; set aside for 30 minutes. Meanwhile prepare the filling by blending ingredients. Now pulverize the nori and blend into batter. Beat in 3 tablespoons oil. Except for nori flakes, batter should be smooth.

Cook crêpes as usual, on a greased pan or griddle. Spread 1-1/2 teaspoons filling across each crêpe, and roll.

Hijiki

Bearer of Wealth & Beauty

There is an old Japanese saying that anyone who eats hijiki will get rich, and it is common knowledge in Japan that hijiki is a boon for the beauty-conscious. Wealth plus beauty may sound like an extravagant promise for a humble sea plant, yet the normalizing effect hijiki has on the blood sugar level is reflected in good coloring and the overall radiance of good health. It also thickens and adds luster and resilience to hair (as do arame, wakame, and nori). In *Healing Ourselves,* Naboru Muramoto prescribes hijiki and wakame for hair loss.

Hijiki is also a treasure for dieters. Dried hijiki absorbs water and expands by as much as four or five times, making it a filling meal adjunct with a calorie count that is almost nonexistent. Furthermore, a daily portion of hijiki insures a full complement of vitamins A, B_1, B_2, and nicotinic acid, along with the sea vegetable's usual concentration of minerals. Hijiki is an excellent source of calcium: an average-size portion (100 grams) provides 1,400 micrograms of calcium, fourteen times the amount in a glass of cow's milk. The bone-building calcium soothes nerves and keeps hormones functioning; since it is quickly depleted by bodily processes, it must be replenished daily.

Hizikia fusiforme (or sometimes *Cystophyllum fusiforme* or *Turbonaria fusiforme*) spreads a brown carpet over rocks near the low-water mark along the Japanese coast, from Hokkaido in the north to Kyushu in the south. Hijiki is classified as a member of the order *Fucales* and thus is related to the rockweed kelps (*Fucus*). The Japanese also classify it with the Sargassum, which, with its dainty blades radiating from long branches, it closely resembles. These branches extend along the rock surfaces and sea bottom almost like rhizomes (horizontal stem-root systems). Erect frondage, called "bush," stands straight up from the horizontal branches. Each "bush" burgeons to a length of 40 inches and weighs a kilogram or more. Plants in more protected environs may get through the summer without suffering too severe defoliation and grow to be 6 to 8 feet long.

Autumn is budding season, and new cells in the horizontal branches grow during the winter. With the spring flood tides, the plants release a myriad of eggs and spermatazoids, which mate and disperse the population. The horizontal branches are perennial and continue to generate new tissue even after the spent receptacles and erect fronds have been torn away by rough water during the summer months. Hijiki attains its maximum size during April and can be harvested between January and May; the tenderest, best-tasting buds are reaped in the colder months of January and February. Harvested plants are allowed to dry in the sun, then are boiled (to make them soft), and again dried thoroughly. The brown plants become black when they reabsorb the concentrated pigment released by the boiling water.

It is always a delight to freshen hijiki and watch the dainty flowerlets come to life in the water. One-quarter cup of dried hijiki will swell enough in water to fill a whole cup. Ten or fifteen minutes' soaking is usually sufficient for the hijiki to regain a slightly crunchy, vegetable softness; drained, it can be used to add a sea flavor to salads, wok dishes, stews, and soups. It makes a great stuffing for squashes and similar vegetables, empanadas, or grape leaves, and adds a protein and flavor boost to rice.

Broccoli & Hijiki Soup   Serves 4

    4 tablespoons corn oil
    1 large onion, sliced
    1-1/2 pounds broccoli stems, chopped
    5 cups water
    3 bay leaves
    1 tablespoon celery seeds
    1 teaspoon kelp powder
    1 tablespoon basil
    1/2 cup dried hijiki, rinsed and drained
    2 tablespoons natural soy sauce (shoyu)
    A dash of lemon juice (optional)

In a deep soup pot (preferably enamel) heat oil and sauté onion and broccoli for 2 or 3 minutes. Add the next five ingredients and simmer, covered, for 1 to 2 hours, until broccoli stems are very soft. Add hijiki and simmer for 1 hour more. Season to taste with soy sauce and, if used, lemon juice. Serve topped with an additional sprinkling of celery seeds, if desired.

Hijiki with Black Beans   Serves 4

    3 cups water
    1/2 pound (1 cup) black beans, rinsed
    2/3 cup dried hijiki, freshened in water and drained
    5 tablespoons natural soy sauce (shoyu)
    2 tablespoons honey
    1/2 teaspoon each, oregano and basil
    1/2 teaspoon savory
    A handful of parsley, chopped

In a large (preferably ceramic) pot, bring the water to a boil, drop in the beans, cover, and return to a boil. Turn off heat and let stand for 1 hour or more. Now, simmer the beans for 1-1/2 hours; add hijiki, half the soy sauce, honey, oregano and basil. Continue to simmer until the beans are tender, about 1 to 2 hours. Stir in the remaining soy sauce and savory, top with parsley, and serve.

Hijiki Potato Salad

Serves 4–5

Seasoning Mixture:

- 1/2 cup water
- 3 tablespoons natural soy sauce (shoyu)
- 1-1/2 tablespoons sake
- 2 tablespoons honey

- 1/2 cup dried hijiki, freshened in water and drained
- 8 medium potatoes, peeled and boiled
- 1/3 teaspoon kelp powder
- 2 tablespoons vinegar
- 1 scallion, chopped
- 5–6 tablespoons mayonnaise (or mayonnaise substitute)
- Lettuce leaves
- 1 tomato, sliced in wedges

Prepare seasoning mixture in a small saucepan and bring almost to a boil. Turn down heat, add hijiki, and heat (do not boil) for 5 minutes. Remove pan from stove and allow hijiki to cool in the broth.

When potatoes have cooled to room temperature, cut into 1/2-inch-thick rounds. Sprinkle with kelp powder and vinegar, then set aside.

Remove cooled hijiki from the broth, and drain. Toss with potatoes, scallion, and mayonnaise. Arrange hijiki–potato mixture on a bed of lettuce leaves, and top with tomato wedges.

Curried Hijiki and Apple Salad

Serves 4

- 2 tablespoons vegetable oil, preferably peanut or sesame
- 1 teaspoon black mustard seeds
- 1 clove garlic, minced
- 1 slice fresh ginger root, slivered (1/4–1/2 teaspoon)
- 1/2 cup chopped onions
- 1/2 to 1 teaspoon curry powder
- 1/4 cup coconut shreds, fresh or dried
- 8 fresh dates
- 2 tablespoons fresh currants or chopped raisins
- 1/2 cup dried hijiki, freshened in cold water for 15 minutes and drained
- 1 lime
- 2 stalks celery, cut crosswise into thin slices
- 1 pound apples, peeled (if desired), cored, and cut into bite-size pieces
- 3/4 to 1 cup mayonnaise (or mayonnaise substitute)
- A sprinkling of freshly grated nutmeg and black pepper
- A few sprigs of fresh coriander, leaves stripped from stems (optional)

Heat oil in a wok and sprinkle in mustard seeds. When seeds begin to pop, quickly add garlic, ginger, and onions, stir-frying for 2 to 3 minutes. Stir in curry powder and fry 1 minute more. Now stir in the next four ingredients and continue stir-frying until all liquid is absorbed or evaporated.

Transfer mixture to a large glass or ceramic bowl, and toss with a few drops of lime juice. Fold celery, apples, and mayonnaise into hijiki mixture and season to taste with lime juice. Serve sprinkled with nutmeg and garnished with coriander.

Hijiki and Wakame Salad   Serves 4

    1 tablespoon sake
    1 tablespoon natural soy sauce (shoyu)
    3 tablespoons water
    1/4 cup dried hijiki, freshened in water and drained
    1 teaspoon rice vinegar
    1 cup dried wakame, cut into 1-inch lengths, freshened in water, and drained
    1 tablespoon lemon or lime juice
    2–4 tablespoons chopped celery leaves or parsley

In a small saucepan, combine sake, soy sauce, and water. Add hijiki and simmer for about 3 minutes. Meanwhile, stir vinegar into wakame and let stand for 2 or 3 minutes. In a salad bowl, combine hijiki, wakame, and juice. Garnish with celery leaves or parsley.

Tomato "Flowers" with Vinegared Miso   Serves 4

    1–2 tablespoons vegetable oil
    1/3 cup dried hijiki, freshened in water and drained
    1 tablespoon natural soy sauce (shoyu)

Dressing:

    5 tablespoons miso
    1 tablespoon honey
    1/4 cup sake
    1/4 cup water
    1/8 teaspoon kelp powder
    1-1/2 tablespoons rice vinegar

    4 medium tomatoes
    1 teaspoon slivered fresh ginger
    2 cups alfalfa sprouts
    1 cup grated cucumber

Heat oil in a pan and sauté hijiki for 1 minute. Add soy sauce and reduce heat to low; simmer, uncovered, for 1 minute. Remove hijiki from pan and allow to cool to room temperature. Meanwhile, combine the first five ingredients of the dressing in a saucepan over low heat for 1 minute, mixing well to dissolve miso thoroughly. Remove from heat, and allow to cool to room temperature. Stir in vinegar, and set aside.

    Slice the tomatoes into eighths, almost but not quite all the way through, as shown. Arrange the tomatoes, open like flowers, on a large serving plate or in individual dishes. Mix the ginger and hijiki and distribute at the center of each "flower." Make a bed for the tomatoes of alfalfa sprouts with grated cucumber sprinkled on top. Trail the dressing over the "arrangement" and serve.

Sweet Vinegar-Ginger and Hijiki Rice

Serves 4

2 cups brown rice
3 cups water
6 inches dried kombu

Sweet Vinegar-Ginger Marinade:

1 teaspoon rice vinegar
1/2 teaspoon sake
1/4 teaspoon honey
1-1/2 tablespoons roughly chopped fresh ginger

1/4 cup vegetable oil
2 cakes tofu, each pressed, drained well and cut into
    8 or 10 pieces
2 tablespoons sake
2 tablespoons honey
2 tablespoons natural soy sauce (shoyu)
3 tablespoons water
1/4 cup dried hijiki, freshened in water and drained

Cook the rice as usual, with kombu on the bottom of the pot. While rice is cooking, combine marinade ingredients and marinate ginger for 30 minutes.

Meanwhile, in a wok or small enamel pan, heat 2/3 of the oil and fry tofu until golden on both sides. Drain tofu well on absorbent paper and cut into thin slices.

In another pan, combine remaining oil with sake, honey, soy sauce, and water over medium heat. Add hijiki and cook for 8 to 10 minutes. Remove hijiki from sauce and stir it into the hot cooked rice, along with fried tofu, mixing thoroughly. Garnish with marinated ginger and serve hot.

Sautéed Hijiki with Sesame Seeds

Serves 4

Its high concentration of sea minerals gives fresh hijiki a somewhat "fishy" scent, which can be satisfactorily diluted by simmering for two or three minutes in a flavored blend of water, sake, and soy sauce. It is good to sauté hijiki in a little oil (preferably cold-pressed, or dark sesame), thereby releasing the oil-soluble vitamins carotin and vitamin E, as well as the flavor. In addition, the piquancy of the seaweed is well complemented by the flavor of the oil.

2-1/2 to 3 tablespoons dark sesame oil
1/2 cup dried hijiki, freshened in water and drained
3 tablespoons natural soy sauce (shoyu)
2 tablespoons sake
1 large tablespoon honey
1 tablespoon sesame seeds

Heat oil in a skillet or enamel pan. Add hijiki and sauté for a few seconds; add remainder of ingredients and simmer for 4 to 5 minutes. Serve hot, or cool to room temperature.

Color Show

Serves 4

   5 tablespoons oil
   2 cakes tofu, each pressed, drained, and cut into 8 or 10 pieces
   1/2 cup dried hijiki, freshened in water and drained
   2 medium carrots, slivered
   6 dried Japanese mushrooms, freshened until soft, in 1 cup water and drained (reserving the soaking water), trimmed, and slivered
   2 tablespoons sake
   5 tablespoons natural soy sauce (shoyu)
   3 tablespoons honey
   1/4 cup chopped parsley

Heat oil in a deep enamel pan or skillet. Add tofu and fry until golden on both sides. Drain fried tofu on absorbent paper, then cut into thin strips. Reheat remaining oil and lightly sauté hijiki, carrots, and mushrooms. Add tofu and mushroom soaking water, and simmer, covered for 3 minutes. Season with sake, soy sauce, and honey, and continue to simmer until almost all liquid is absorbed or evaporated (about 15 minutes). Serve garnished with parsley.

Spicy Szechuan Tofu with Hijiki

Serves 4

Sauce:

   1 cup water
   1 tablespoon kudzu powder (or arrowroot or potato starch)
   4 tablespoons natural soy sauce (shoyu)
   1/2 teaspoon anise seed

   4 tablespoons oil (combination of safflower and dark sesame)
   1 clove garlic, minced
   1 tablespoon chopped ginger
   1/2 cup dried hijiki, freshened in water and drained
   3 cakes tofu, each pressed, drained and cut into 8 pieces
   1 scallion, chopped
   A sprinkling of cayenne or chili powder

Prepare sauce and set aside. Heat oil in a wok or heavy skillet, and saute garlic and ginger. Add hijiki and sauté for 1 minute, until the hijiki absorbs most of the oil. Add tofu and sauté for about 30 seconds. Now, add sauce and, stirring constantly, cook over medium heat until it thickens. Add scallion, season to taste with cayenne, and remove from heat.

Mexican Fiesta

Serves 3–4

This dish is delicious with tortillas and guacamole.

    4–5 tablespoons corn oil
    2 cloves garlic, chopped
    2 onions, chopped
    1 or more jalapeña peppers, chopped
    1/2 cup dried hijiki, freshened in water and drained
    4 medium or 2 large squash, diced
    1 or 2 tomatoes, cut in eighths
    1/2 cup water
    A sprinkling of kelp powder
    A pinch of oregano or other seasoning

Heat oil in a skillet and sauté garlic, onion, and hot pepper, then hijiki and squash. Now add tomatoes and water, and simmer, covered, until the tomatoes are soft (about 10 minutes). Season to taste with kelp powder and oregano.

Hijiki Empanadas

Serves 4–5

    1/2 cup dried hijiki
    1 tablespoon olive oil
    1 clove garlic, minced
    1 small onion, chopped
    1 small sweet bell pepper, chopped
    Minced hot chili pepper and/or cayenne to taste
    1/2 teaspoon each, ground cumin, basil, thyme
    1/4 teaspoon ground allspice
    Juice of 1/2 fresh lime
    Freshly ground black pepper
    1 tablespoon currants, freshened in water (or freshened and
        chopped raisins)
    12–16 ripe green olives, pitted and chopped
    12 (6-inch) pastry rounds

Cover hijiki with 2 cups water and set aside, adding more water from time to time, as needed. Heat oil, and sauté garlic, onion, bell pepper, and spices until onion is transparent. Add undrained hijiki with not more than 1/3 cup soaking water. Increase heat slightly and, stirring frequently, mix all ingredients thoroughly, cooking off most of the liquid. Turn off heat and stir in lime juice, black pepper, currants, and olives.

Preheat oven to 375°. Use the pastry rounds and hijiki-vegetable mixture to prepare 12 turnovers or 24 smaller hors d'oeuvres. Bake for 25 to 30 minutes, or until golden.

Baked Stuffed Pumpkin
Serves 10–12

This festive dish will be the *pièce de resistance* at a vegetarian Thanksgiving Day table, and it's simple to prepare.

- 2–3 tablespoons vegetable oil
- 1 small onion, slivered
- 1 clove garlic, minced
- 1/2 teaspoon each sage, thyme, paprika
- A sprinkling of freshly grated nutmeg
- 1 teaspoon natural soy sauce (shoyu)
- 1 stalk fresh fennel, cut crosswise into narrow slices (or substitute celery and add 1/4 teaspoon crushed fennel seed to spices listed above)
- 1 cup sliced mushrooms
- 12 chestnut meats, chopped (about 3/4 cup)
- 1/4 cup dried hijiki, freshened and drained
- 1/3 cup dried wild rice, slightly undercooked (if unavailable, double the amount of hijiki used)
- 1 cup brown rice, slightly undercooked
- 1/4 cup finely chopped fresh parsley
- 1 teaspoon sea salt, approximately
- 1 (5-pound) pumpkin with a lid cut into the top, seeds and membranes removed

Heat 2 tablespoons oil in a large skillet and sauté onion, garlic, herbs, and spices for about 3 minutes. Add soy sauce and the next four ingredients, one at a time, sautéing each for 1 minute. Remove pan from heat and stir in both varieties of rice and parsley.

Preheat oven to 350°. Lightly salt pumpkin cavity and fill with the hijiki–rice stuffing. Replace lid and secure with toothpicks. Coat surface of pumpkin with oil. Bake for 1-1/4 hours, or until tender.

Variation: Use the stuffing to fill 10–12 large sweet bell peppers; bake at 375° for 20 to 25 minutes.

Shepherd's Pie
Serves 4–6

- 2 tablespoons vegetable oil
- 1 clove garlic, minced
- 1 onion, chopped
- 1 sweet bell pepper, diced
- 1/2 pound mushrooms, chopped
- 1/2 teaspoon each thyme and basil
- 1/4 cup dried hijiki, freshened in water to cover
- 2 cups cooked kidney beans, seasoned with 1/2 teaspoon each kelp powder and savory herb and 1/4 teaspoon salt
- 2-1/2 cups baked yam purée or winter squash puree
- 1 (9-inch) prebaked pie crust
- A sprinkling of cayenne, paprika, and nutmeg

Heat oil and sauté next five ingredients for 5 minutes, or until onion is transparent. Add hijiki, mixing well to coat it evenly with oil, and sauté for 2 to 3 minutes more before removing from heat. Preheat oven to 350°. Layer vegetables on crust, alternating layers of purée with layers of beans and hijiki; sprinkle topmost layer of purée with spices. Bake for 20 to 25 minutes, until crust is golden brown. Allow to set for 5 to 10 minutes before slicing.

Tahini Quiche

Serves 4–6

Here is an eggless, milkless quiche with a full rich texture.

Crust:

1-1/2 cups whole wheat pastry flour
1/4 teaspoon sea salt
1/4 cup corn oil
1/4 cup cold water

Filling:

2 tablespoons oil
1 clove garlic, minced
1 onion, chopped
1/4 cup dried hijiki, freshened in water and drained
1/4 teaspoon cumin powder
1/4 teaspoon basil
A sprinkling of cayenne
A sprinkling of nutmeg
1 cup spinach, roughly chopped
1/2 cup tahini
3 tablespoons kudzu powder (or arrowroot or potato starch)
1 tablespoon honey
1-1/2 cups water
1 cake tofu, mashed with a fork (optional)

Preheat oven to 350°.

Crust: Combine flour and salt, mixing well. Cut in corn oil with a fork or fingertips until oil is evenly distributed and dough is crumbly. Stir water in quickly, mixing thoroughly. (The dough should be pliant but not sticky. Add more water if dough is too dry.) Form dough into a large ball, flatten, and turn into a greased 8-inch baking dish or deep pie plate, using fingers to spread the dough and form 1-1/2-inch-high sides. Prebake for 5 minutes at 350°.

Filling: Heat oil in a skillet or wok, and sauté garlic, onion, hijiki and spices. Fill the pastry shell with mixture, and layer the spinach on top. Combine tahini, kudzu, honey, water, and, if used, tofu; pour mixture over vegetables. Bake at 350° for 30 minutes or until firm.

Sweet Potato and Hijiki Balls

Serves 4

3 medium sweet potatoes, boiled and mashed
1/2 cup dried hijiki, freshened in water and drained
2 tablespoons honey
1/2 cup brown rice flour
4–6 tablespoons vegetable oil
3 sheets of nori, crisped
1 cup snow peas (optional)
2 green peppers (optional)

Cut hijiki and honey, then rice flour, into mashed potatoes, mixing thoroughly. Form 12 half-dollar-size patties and fry in vegetable oil until golden. Drain well on absorbent paper.

Cut each sheet of nori into quarters, and cut each quarter into halves. Sandwich each patty between two pieces of nori.

Sauté green pepper and snow peas, if used, in the remaining oil for 30 seconds, or until their color deepens. Serve as garnish for hijiki balls.

Hijiki and Carrots with Creamy "White Sauce"

Serves 4

Seasoning Mixture:

- 2 cups water
- 3 teaspoons natural soy sauce (shoyu)
- 1/4 teaspoon kelp powder
- 1 teaspoon honey

2 or 3 medium carrots, cut into 1-inch-long slivers
1/4 cup dried hijiki, freshened in water and drained

White Sauce:

- 1/2 cake tofu
- 2 tablespoons tahini
- 1-1/2 tablespoons honey
- 1/3 teaspoon kelp powder
- 1 teaspoon natural soy sauce (shoyu)

Prepare seasoning mixture in a deep saucepan and bring almost to a boil. Add carrots and hijiki, and bring almost to a boil again. Then reduce heat and simmer, covered, until the carrots are tender (5 to 10 minutes).

To prepare white sauce, drop tofu into boiling water to cover for 30 seconds; drain well and mash with a fork. Mix in remaining ingredients, and thin with liquid from the simmering seasoning mixture. Pour over vegetables and serve.

Vegetarian Stir-Fried Mock Crab Meat

Serves 4

- 8 dried Japanese mushrooms
- 1/2 cup vegetable oil
- 1/3 cup dried hijiki, freshened in water and drained
- 1 cup cooked carrot, puréed (about 1 large carrot)
- 1 cup cooked potato, puréed (about 3 small potatoes)
- 1 cup slivered bamboo shoots
- 1/4 pound snow peas, slivered
- 2 tablespoons honey
- 1/4 teaspoon kelp powder
- 2 teaspoons rice vinegar

Soak mushrooms in hot water to cover until soft (about 15 minutes). Sliver softened mushrooms and set them aside. In a wok or enamel pan, heat oil and stir-fry hijiki, carrot, and potato until crispy (about 4 minutes). Add bamboo shoots, mushroom slivers, and snow peas, stir-frying for 2 minutes more. Add honey and kelp powder, mixing well. Now quickly stir in vinegar. Serve hot.

Agar

Ahnfeltia

Sea Vegetable Jello

Three hundred years ago in Japan, on a cold winter night, innkeeper Minoya Tarozaemon was doing everything within his power to make the visiting lord of an important Kyushu fief comfortable during his stopover in Fushimi, near Kyoto. For dinner Minoya served a gourmet meal featuring the best local cuisine. One boiled dish included the highly favored sea vegetable *tengusa*. The meal was too extensive for even the kitchen help to finish off, and some of the leftover tengusa was thrown out on the ground. The tengusa froze, then dehydrated naturally, becoming light and dry. Several days later Minoya discovered the new substance and was intrigued by it. Being an ingenious sort, he tried melting it in boiling water and found that upon cooling it turned into a white substance that had, in combination with other foods, a luscious taste. He then refined the process and sold it to a firm in Edo (now Tokyo). The processing of tengusa to produce *kanten* has changed very little since that time.

So the story goes in Japan. The Chinese believe that they were the ones who taught the Japanese how to freeze-dry species of *Gelidium* into translucent bars of kanten. Determining who taught whom is difficult, as tengusa—"the grass of heaven"—is mentioned in very early texts in both countries. In fact, the *Chi Han,* from 300 A.D., recommends agar to control insect pests.

Agar is a Malay word which means "jelly." Basically, agar is a complex polysaccharide, or starch, related to cellulose and found in the cell walls of agarophytes (agar-yielding species of red algae). Many peoples throughout the world have long recognized agar's gelling properties and used it to make sweetmeats. The Japanese favor *Gelidium amansii* as a source for kanten, but with the global burgeoning of the agar industry since World War II, a wide number of agarophytes have been conscripted for commercial use: species of *Pterocladia, Gracilaria, Acanthopeltis, Ahnfeltia, Gigartina,* and other *Gelidia. Pterocladia* and *Gelidium*, both common in the British Isles, are most widely harvested for agar today.

Most of the agarophytes are still harvested by fairly primitive methods. Harvesters collect them by hand or dive for them; sometimes they are raked in from boats. There is no specific optimal season for harvesting, but photosynthesis does reach a peak at certain times (May to July for *Gelidium*), improving the agar's quality.

Agarophytes flourish everywhere in the world, down to depths of 200 feet. Flat and horny strands of parsley- or fern-like fronds, waving in browns, reds, and purples, form a luxuriant carpet 1–4 inches thick on the coastal rocks. Each type has its peculiarities. Its delicate filamentous branching notwithstanding, *Ahnfeltia* is the stiffest of all red algae and washes ashore in bundles. *Gigartina* is the largest genus, encompassing some ninety species, including some of the largest red algae: their feather-shaped blades, the color of red cabbage, may extend 3 feet from their rocky mounts. *Gracilaria* accumulate in growing masses up to several feet thick; their weight may increase tenfold as they drift along the coasts. *Gelidium* is often accompanied by a mascot: goldfish share its habitat on the seaward edges of rocky reefs.

A prominent Buddhist priest in Japan once commented that kanten is "the perfect Buddhist food." Vegetarian dishes such as "Takigawa Dofu" and "Mizu Yokan" continue to please tea ceremony guests and priests with gourmet palates. Agar has a way of emphasizing the natural sweetness of fruits and vegetables, without interfering with their taste. It can be molded or cut into cubes and slices and arranged decoratively, turning ordinary desserts into *pièces de resistance*. Agar also makes delicious pies which require no eggs or dairy products. Ideal for dieters, agar (like fiber) lends bulk without calories. It provides appreciable quantities of calcium, iron, phosphorus, and vitamins A, B_1, B_6, B_{12}, biotin, C, D and K. It not only aids intestinal action, promoting digestion, but bonds with toxic and even radioactive wastes and carries them out of the body (see *Nature,* December 25, 1965; see also *Prevention,* August 1972). Furthermore, it sets at room temperature (though refrigeration will speed up the process).

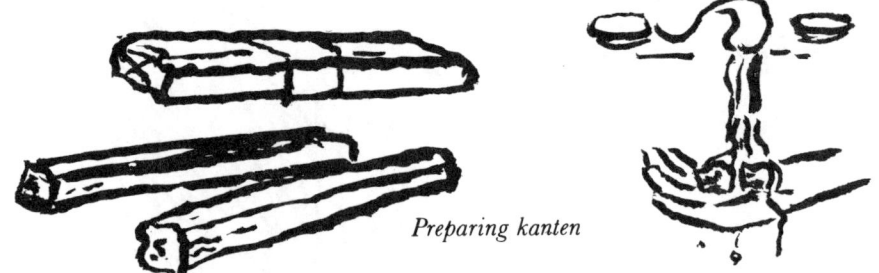

Preparing kanten

Agar is available in various forms: featherlight bars of kanten, strands, powder, and flakes. Agars vary in their gelling properties (it may take only 2-1/2 tablespoons of the lighter-colored flakes and up to 4 tablespoons of the darker brownish variety to gel 4 cups of water, juice, or milk, or 3 cups of heavier liquids such as purées or soy milk). Agar will not set in the presence of acetic acid (found in distilled and wine vinegars); thus, it is better to use lemon juice (citric acid) or apple or malt vinegars (malic acid) when making aspics or savories. Nor will it set with the large quantities of oxalic acid found in chocolate, rhubarb, and spinach; for these foods, Irish moss must be substituted.

To prepare, kanten bars need only be broken into several pieces, washed, wrung out (by squeezing with the hands), and soaked in water for at least 30 minutes. Strands should likewise be washed, but they will soften in water very quickly, in 1 or 2 minutes. To dissolve, the liquid should be heated slowly, almost to a boil, and stirred constantly; then the desired flavoring may be added. Kanten bars may have impurities which can be removed by straining the dissolved liquid through cheesecloth stretched over a bowl or measuring cup.

The suggestions presented here are just an introductory guide to incorporating agar into daily cooking. The possibilities are endless.

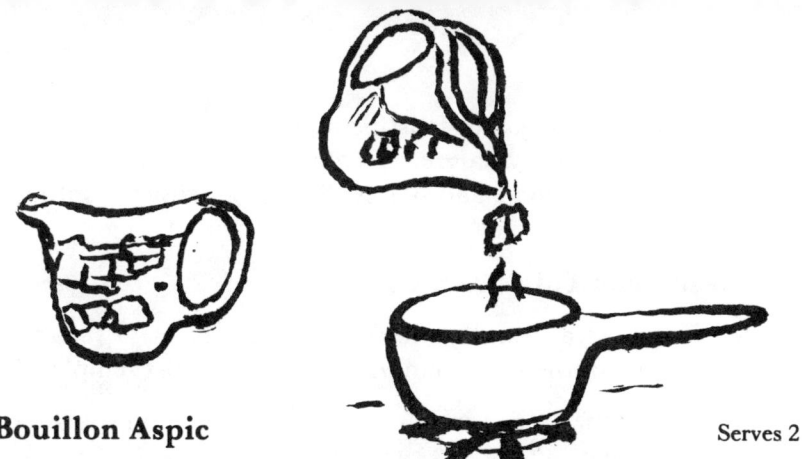

Bouillon Aspic

Serves 2

An especially cooling and nutritious way to enjoy potassium-rich vegetable broths and bouillons, including Kombu Dashi.

- 1-1/2 tablespoons agar flakes (or 1/2 bar kanten, broken into 2 or 3 pieces, washed and squeezed)
- 2-1/2 cups homemade vegetable broth prepared from selected trimmings, kelp powder, and garlic (or 2-1/2 cups water and 3 teaspoons instant natural vegetable bouillon powder)
- 2 teaspoons natural soy sauce (shoyu)
- 2 teaspoons lemon juice
- A pinch of cayenne
- 2 slices of fresh lemon
- Sprigs of fresh green herbs or watercress

Soak agar in the broth for 30 minutes or more to soften; bring almost to a boil over medium heat, stirring frequently, until agar dissolves. (If using bouillon powder, soak agar, then bring almost to a boil in water; remove from heat and stir in bouillon powder.) Add lemon juice, soy sauce, and cayenne. Pour into premoistened individual glass serving bowls. When set, stir once, if desired. Serve garnished with lemon slices and fresh herbs.

Lemon Bouillon Aspic

Serves 4

- 2-1/2 to 3-1/2 tablespoons agar flakes (or 1 bar kanten, broken into 6–8 pieces, washed and squeezed)
- 4 cups water
- 2-3 tablespoons lemon juice
- 1/2 teaspoon crushed lemon verbena leaves
- 1/2 teaspoon marjoram
- A pinch of allspice powder
- A dash of sea salt
- A dash of red or black pepper (optional)
- Garnishes, as desired

Soak agar in the water for 30 minutes or more, then heat and dissolve as in Bouillon Aspic. After removing dissolved agar from heat, stir in remaining ingredients; strain through cheesecloth, if desired. Pour mixture into premoistened molds and allow to set. Cube and use as an addition to salads.

Layered Avocado & Vegetable Mold
Serves 4

Though very colorful and intricate in appearance, this succulent salad is not difficult to make.

> 2-1/2 to 3-1/2 tablespoons agar flakes (or 1 bar kanten, broken into 6–8 pieces, washed and squeezed)
> 3 cups water
> 1 teaspoon lemon juice
> A pinch of kelp powder
> 1 large carrot
> 1 small green pepper cut in strips
> 1 avocado, fully ripened
> 1 tablespoon finely grated onion
> 1/3 cup pepitas or lightly toasted sunflower seeds
> 1/2 teaspoon cumin powder
> 1 cup homemade lemon vinaigrette salad dressing (see Wakame Vinaigrette)
> 1/2 bunch watercress
> 1/2 cup (or more) fresh sprouts

In a 2-quart saucepan, immerse agar in water to soften. Over high heat bring agar almost to a boil, stirring occasionally, until it dissolves. Remove pan from heat and stir in lemon juice and kelp. Set pan aside and allow agar to cool slightly.

Grate carrot and divide among four 2-cup glass salad bowls (or use individual fancy molds) slightly moistened with water to prevent sticking. Gradually pour in enough agar liquid almost to cover the carrot, and allow to cool and set partially. Distribute the green pepper strips atop the carrot layers and almost cover with agar. (If agar has solidified on the stove, gently reheat to liquify.) Now mash avocado well, and blend in onion and remaining agar liquid, beating until smooth. Pour mixture over green pepper layer and refrigerate for 30 minutes.

Just before serving, pan-toast pepitas and cumin, then crush together and stir into vinaigrette. Unmold salads onto beds of watercress arranged on salad plates. Serve topped with sprouts and dressing.

Japanese Agar Tofu (*Takigawa-Dofu*)
Serves 4

Named for an area near Kyoto, this dish is a favorite Japanese delicacy served in the tea ceremony and in Buddhist temples.

> 1 bar kanten, broken into 5–6 pieces, rinsed, and squeezed
> 2 cups water
> 2 cakes tofu, mashed with a fork
> 1/2 teaspoon kelp powder
> 1 tablespoon honey

Sauce:

> 1 tablespoon mustard powder
> 2–3 tablespoons natural soy sauce (shoyu)

Immerse kanten in water for at least 30 minutes. Combine tofu, kelp powder, and honey, mixing thoroughly, and set aside.

Transfer kanten and water to a saucepan and, over medium heat, bring almost to a boil, stirring constantly until kanten dissolves. Strain kanten mixture through a cheesecloth placed over a bowl or measuring cup, then combine with tofu mixture in an electric blender; blend well. Turn into a moistened glass baking dish or mold, and refrigerate until firm (about 30 minutes). Cut into slices and arrange on a serving platter. Serve with dipping sauce.

Variation: Parboiled vegetable pieces—carrots, cauliflower, broccoli, etc.—may be folded into the agar after it has partially set.

Japanese Aspic with Ginger Sauce

Serves 4

1 bar kanten, broken into 5–6 pieces, rinsed and squeezed (or 2-1/2 to 3-1/2 tablespoons agar flakes)
2 cups water
1/3 cup wakame, cut into 1/2-inch-long pieces, freshened in water, and drained
1 teaspoon rice vinegar
1/4 teaspoon kelp powder
1 tablespoon natural soy sauce (shoyu)
1 tablespoon sake
1/2 teaspoon honey
1/2 cup grated carrot

Ginger Sauce:

1/2 cup kombu broth (or water with 1/4 teaspoon kelp powder)
1-1/2 tablespoons natural soy sauce (shoyu)
1-1/2 tablespoons sake
1/2 teaspoon honey
1/2 teaspoon grated fresh ginger

A few sprigs of parsley

Gigartina

Immerse kanten in the water for at least 30 minutes. In a separate bowl, sprinkle vinegar over wakame and set aside.

Turn the kanten and soaking water into a saucepan and bring almost to a boil over medium heat, stirring until kanten is completely dissolved. Remove pan from heat and strain contents through a strip of cheesecloth stretched over a mixing bowl. Stir kelp powder, soy sauce, sake and honey into kanten mixture. When mixture has cooled to body temperature, stir in wakame and carrot; turn into a moistened glass loaf pan or jello mold. Allow to cool at room temperature until fully set.

Simmer the sauce ingredients over low heat for 1 minute, or until heated through, then allow to cool to room temperature. When ready to serve, slice the jelled kanten and arrange the slices attractively in a serving bowl. Serve the aspic topped with sauce and garnished with parsley sprigs.

Marinated Agar Strands and Spinach Salad

Serves 4

Dressing:

- 2 tablespoons wine vinegar
- 2 tablespoons safflower oil
- 3–4 tablespoons mayonnaise (or mayonnaise substitute)

- 1 cup 2- to 3-inch strands agar, freshened in warm water and squeezed to dry
- 2 cups spinach, washed and trimmed

In a salad bowl, combine dressing ingredients, mixing well: marinate agar strands in mixture for 4 to 5 minutes. Now toss spinach with agar strands and dressing, and serve.

Molasses Spice Aspic

Serves 4

- 2 tablespoons agar flakes (or 1/2 bar kanten, broken into 2 or 3 pieces, washed and squeezed)
- 2-1/2 cups water
- A small piece of sassafras bark or a 1-inch-long cinnamon stick
- 1 teaspoon freshly grated orange peel
- 2 tablespoons molasses
- 1 tablespoon honey
- 1 tablespoon lemon juice

Soak agar in water for 30 minutes or more. Add sassafras or cinnamon and dissolve over low heat, as in Bouillon Aspic. After removing dissolved agar from heat, stir in remaining ingredients. Pour into a premoistened mold and allow to set.

Molded Fruit & Nuts

Serves 4–6

- 2-1/2 to 3-1/2 tablespoons agar flakes (or 1 bar kanten, broken into 6–8 pieces, washed and squeezed)
- 4 cups water (or substitute 2 cups apple, grape, or orange juice for 2 cups of the water)
- 1–4 tablespoons honey
- Juice of 1/2 lemon
- A pinch of kelp powder (optional)
- 1 cup fresh fruit, cut into bite-size pieces
- 1/2 cup chopped nuts

In a 2-quart saucepan, immerse the agar in 2 cups water until softened, then heat over medium heat until agar dissolves. Reduce heat to low, and stir in honey and 2 cups water. Remove pan from heat, then stir in lemon juice and, if used, kelp powder; set mixture aside to cool slightly.

Arrange the fruit in a mold or serving dish moistened with water to prevent sticking. Slowly pour in sweetened agar and allow to gel partially; top with nuts, and allow to set thoroughly.

Baked Spiced Pears in Cider Aspic

Serves 4–8

- 2-1/2 to 3-1/2 tablespoons agar flakes (or 1 bar kanten, broken into 6–8 pieces, washed, and squeezed)
- 3 cups water
- A 1-inch-long cinnamon stick
- 2 slices fresh ginger root
- 1 cup apple cider (hard cider is fine)
- 1/4 cup maple syrup (or 2 tablespoons molasses)
- Juice of 1/2 lemon
- 1/4 cup walnut pieces
- 1/4 cup currants or raisins
- 1 teaspoon slivers of fresh lemon rind
- 1 teaspoon light-tasting vegetable oil
- 4 large fresh pears
- 1 tablespoon whole cloves (approximately 24)

Combine agar, cinnamon, and water in a medium saucepan, and allow agar to soften. Add ginger slices and bring almost to a boil, stirring occasionally, until agar dissolves. Reduce heat to low, add cider and syrup, and simmer until heated through. Remove pan from heat and stir in lemon juice, walnuts, currants, and lemon rind; set aside.

Preheat oven to 350°. Halve and core pears, and arrange halves, peel up, in a single layer in an oiled glass or enamel baking dish. Push a few cloves in each half, then pour on cider aspic to cover. Bake for 45 minutes, basting once or twice.

Puréed Pear Jello

Serves 4

- 1 bar kanten, broken into 5–6 pieces, rinsed, and squeezed (or 2-1/2 to 3-1/2 tablespoons agar flakes)
- 2-1/2 cups water
- 1 cup pear purée (2 medium pears puréed in blender)
- 1/2 cup apple cider
- 3 tablespoons honey
- A squeeze of lemon
- 1/8 cup dried currants

Immerse kanten in water for at least 30 minutes, then transfer kanten and water to a medium saucepan. Over high heat bring almost to a boil, stirring occasionally, until the kanten dissolves. Remove from heat and strain through cheesecloth into a mixing bowl. Stir in the remaining ingredients, except for the currants. Turn kanten mixture into a moistened mold or glass dish, and allow to cool at room temperature until partially set. Sprinkle in currants and allow to set fully.

Almond Sweet (*Annin Dofu*)   Serves 4–5

Japan's sweet tooth was nurtured by the Chinese, who, during the fourteenth century, introduced from the mainland such delicacies as mandarin oranges (now a big cash crop in Japan), honey, brown sugar, and sweet syrup, thus making Annin Dofu possible. Annin Dofu may have been a later introduction; in any case, it certainly caught on and can now be enjoyed in almost every Chinese restaurant in Japan. This delightful dessert is simple to make and quick to disappear.

    1/2 bar kanten, broken into 2–3 pieces, rinsed, and squeezed
    1 cup water
    1/3 cup honey
    1 cup almond milk (or 6 tablespoons almond butter blended with 1 cup of water)
    1 teaspoon almond extract
    1 11-ounce can mandarin orange slices

Syrup:

    1 cup juice from orange slices
    1/2 cup honey

A squeeze of lemon juice

Immerse kanten in water for at least 30 minutes, then transfer kanten and water to a medium saucepan. Over medium heat, bring almost to a boil, stirring constantly. When kanten has dissolved, stir in honey, then strain mixture through a piece of cheesecloth stretched over a bowl or measuring cup.

In a separate pan, warm almond milk to body temperature; add to dissolved kanten, together with the almond extract, stirring well. Pour mixture into a moistened mold and allow to cool at room temperature, then refrigerate for 20 to 30 minutes, or until set.

In a saucepan, bring syrup ingredients almost to a boil. (If the juice from the orange slices is presweetened, lessen the amount of honey used.) Turn off heat and remove pan from stove. When syrup has cooled to room temperature, stir in lemon juice.

To serve, cut kanten into diamonds, pour on syrup, and top with mandarin orange slices.

Grape Jello Cubes with Fresh Fruit Compote

Serves 4–8

Serve this gel as a shimmering mold surrounded by fruits; or, using a melon-baller, scoop out circles and fold in with the fruits.

> 2-1/2 to 3-1/2 tablespoons agar flakes (or 1 bar kanten, broken into 6–8 pieces, washed, and squeezed)
> 2 cups water
> 2 cups grape juice
> A 1-inch-long cinnamon stick
> 4–6 cups fresh fruit pieces, seasoned with fresh lemon and dusted with ground coriander seed

In a 2-quart saucepan, immerse agar and cinnamon stick in the water to soften. Over high heat bring almost to a boil, stirring occasionally, until agar dissolves. Reduce heat to low, add juice, and gently heat through. Turn off heat, remove cinnamon, and pour into a premoistened mold to set.

Grapefruit Dessert

Serves 4

The Japanese serve this dessert with the grapefruits quartered and nestled on a bed of green leaves adorned with seasonal flowers.

> 2 bars kanten, each broken into 5–6 pieces, rinsed, and squeezed
> 4 cups water
> 2 grapefruits
> 1/2 cup honey

Immerse kanten in water for at least 30 minutes. While kanten is soaking, halve grapefruits, extracting juice and removing and discarding pulp. (Each grapefruit should yield 3/4 to 1 cup juice.) Reserve grapefruit shells for later use.

In a saucepan over medium heat, bring kanten and water almost to a boil, stirring constantly until agar is completely dissolved. Remove pan from heat and strain kanten mixture through a cheesecloth stretched over a bowl. Return strained liquid to the pan, stir in honey, and simmer for about 30 seconds. Now stir in grapefruit juice and simmer for 2 or 3 minutes more, stirring occasionally.

Moisten the insides of the grapefruit shells with water and ladle in the agar mixture. Allow to set. Pour the remainder of the agar mixture into individual dessert dishes or a bowl.

Pumpkin Pie

Makes 2 (9-inch) pies

Crust:

- 3 cups whole wheat pastry flour
- 1/2 teaspoon sea salt
- 1/2 cup corn oil
- 1/2 cup water

- 3 cups soy milk
- 1 cup pumpkin purée
- 6–8 tablespoons agar flakes
- 3/4 cup honey
- 1/2 teaspoon each ginger, cloves, coriander, nutmeg
- 1 teaspoon soy flour
- 2 teaspoons vanilla extract
- 1 teaspoon grated orange or lemon peel (optional)

Preheat oven to 350°. Prepare crust and prebake for 8 to 10 minutes. Combine soy milk, pumpkin, and agar in a saucepan and heat almost to a boil, stirring occasionally, until agar dissolves completely. Turn off heat and stir in remaining ingredients. Bake for 15 to 20 minutes.

Pecan Pie

Yields 1 pie

Filling:

- 1/2 teaspoon liquid lecithin
- 1/3 cup vegetable oil (safflower, light sesame, or peanut)
- 1 cup malt syrup
- 2 tablespoons brown agar flakes (or 1 tablespoon white agar flakes)
- 1/2 cup water
- 1/2 cup date sugar
- 1/2 teaspoon vanilla extract
- 1/8 teaspoon grated nutmeg (fresh, if possible)
- 1/2 teaspoon freshly grated lemon rind
- 1-1/2 to 2 cups pecans or other nutmeats, broken into pieces

1 unbaked, chilled, pricked pie shell

Preheat oven to 425°. Beat lecithin into oil until thoroughly blended. In a mixing bowl, lightly stir oil–lecithin mixture into malt syrup. Combine agar and water, mixing well, then add to bowl, together with the next four ingredients; blend well. Now fold in nuts. Turn into a pie shell. Bake for 15 minutes, then reduce heat to 350° and bake for 20 minutes more. Allow pie to cool thoroughly, for approximately 1 hour, before slicing.

Tofu Fruit Pies

Serves 5–8

The succulent flavors of fruits and berries are preserved by using agar instead of pastry starches to thicken pies. Prepare a single recipe of any fruit mixture for use with the tofu filling, or make a double recipe for straight fruit pies. (See Apple-, Blueberry-, or Strawberry-Agar Toppings.)

"Cheesy" Tofu Filling:

    2 cakes tofu
    2–4 tablespoons honey
    A pinch of mace powder
    1 teaspoon citrus rind (lemon or orange)
    1/2 tablespoon arrowroot starch

1 unbaked, chilled, pricked pie shell

Preheat oven to 375°. In a mixing bowl, beat tofu until it is very smooth; blend in remaining ingredients and set aside for a few minutes, to allow the flavors to marry. Bake in pie shell for 20 to 25 minutes. Allow to cool, then top with a fruit and agar mixture.

Variation: Piecrust: Add 1/2 teaspoon of vanilla and 1/4 teaspoon of cinnamon or simply 1/2 teaspoon almond extract.
    Tofu Filling: Add 2 tablespoons raisins or currants.

Citrus Jam

Makes 1 quart

    1/4 cup citrus fruit juice (grapefruit, orange, or other)
    1/4 cup grape juice
    2-1/2 tablespoons agar flakes
    2 cups finely chopped citrus fruit
    1/2 cup honey
    A pinch of grated orange peel
    A pinch of allspice

Pour juice into a 2-quart heavy enamel saucepan and stir in agar flakes. Bring to a boil, then immediately reduce heat and simmer, stirring constantly, until agar is completely dissolved and the liquid is almost clear. Turn off heat. Stir in fruit, honey, and spices, and return to a boil; cook uncovered, for 1 minute. Pour into sterilized covered glass jars and store in a cool place.

Strawberry Jam

Makes approximately 2/3 quart

    2-1/2 tablespoons agar flakes
    1/2 cup apple juice
    2 cups mashed strawberries
    1 cup honey
    1 teaspoon lemon juice

Stir agar into apple juice and bring almost to a boil. Reduce heat and simmer, stirring constantly, until agar is completely dissolved. Stir in fruit, honey, and lemon juice, return to a boil, and cook for 1 minute. Pour into sterilized covered glass jars and store in a cool place.

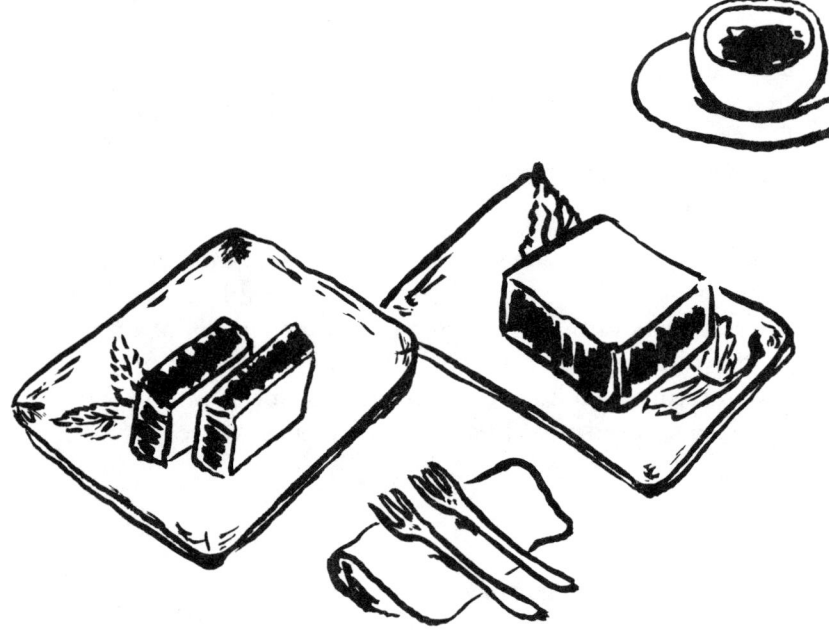

Mizu Yokan, a sweet complement for green tea

Mizu Yokan
Serves 8 or more

Yokan, which may be flavored with chestnuts, sweet beans, persimmons, coffee, green tea, or molasses, as well as azuki beans, resembles guava paste in sweetness and texture. The perfect after-meal sweet, yokan is traditionally served to complement the bitter taste of green tea; it can frequently be found on the menus of Japanese restaurants in America. "*Mizu*" (water) signifies the lighter varieties of this dish, which contain more liquid—and herald the arrival of warm weather.

> 2-1/2 to 3-1/2 tablespoons agar flakes (or 1 bar kanten, broken into 6–8 pieces, washed, and squeezed)
> 2 cups water
> 3/4 cup honey
> 2 cups azuki beans, precooked in water to cover

In a small saucepan, soften agar in water, then bring almost to a boil over high heat, stirring occasionally, until agar dissolves. Reduce heat to low, stir in honey, and simmer agar until heated through. Remove from heat and allow to cool slightly.

Combine agar–honey mixture with beans in an electric blender and blend until smooth. Turn into a premoistened glass mold (preferably square) and refrigerate until set. Serve small pieces.

Blueberry–Agar Topping

Serves 5–8

1-1/2 tablespoons agar flakes
1 cup water
1/4 teaspoon cinnamon powder
1/2 teaspoon ground coriander
1 to 1-1/2 pints blueberries
2 tablespoons honey
1/2 to 1 tablespoon lemon juice

Soak agar and spices in water. When softened, add blueberries and cook over medium heat, stirring occasionally, until agar dissolves and about half of the berries have burst. Reduce heat to low, stir in honey and lemon juice, and remove pan from heat. Cool slightly before gently pouring over the cooled filling. Cool thoroughly before slicing.

Strawberry–Agar Topping

Serves 5–8

1-1/2 tablespoons agar flakes
1-1/2 cups water or strawberry juice
1/8 teaspoon cinnamon powder
Freshly grated nutmeg, to taste
1 tablespoon honey
1/2 to 1 tablespoon lemon juice
1-1/2 pints strawberries

Prepare agar with water and spices according to previous instructions. Over high heat bring almost to a boil, stirring occasionally, until agar dissolves. Reduce heat to low, and stir in honey and lemon juice, then remove from heat. Slice strawberries directly into pot. Allow mixture to cool before spooning onto cooled tofu filling. Slice only when thoroughly cooled and set.

Apple–Agar Topping

Serves 5–8

1-1/2 tablespoons agar flakes
1 cup apple cider
1/4 teaspoon each cinnamon powder, ground cloves, and ground coriander
2 tablespoons honey
1 tablespoon lemon juice
2 pounds apples

Soak agar in cider with spices. Bring almost to a boil over high heat, stirring occasionally, until agar dissolves. Reduce heat to very low, and stir in honey and lemon juice. Now core and slice the apples directly into the pot. (Stir after adding each apple to coat evenly and cook out some of the water in the apples, to ensure uniform gelling.) For a soft texture, cover pot after all apples have been added and simmer for 3 to 7 minutes. Remove from heat and cool slightly before spooning onto tofu filling. Cool pie thoroughly before serving. May be served chilled.

Irish Moss

Like agar, Irish moss (*Chondrus crispus*) has been valued for centuries for its remarkable gelling properties. Known as carrageen (from Carraghean, an Irish coastal town), sea moss provides an extract, carrageenin, with multiple industrial uses. Irish moss accounts for the oldest seaweed industry in the United States, centered in Scituate, Massachusetts, where the seaweed was introduced in the nineteenth century. The Boston Brewery was once a prominent customer: brewers would toss raw whole plants into their vats so that they would bond with impurities and carry them to the bottom.

Harvesters collect Irish moss with long-handled rakes twice a year, during the period when the carrageenin content is highest—from late summer through fall. (Carrageenin constitutes up to 70 percent of the dry moss.) *Chondrus* ranges widely, along the Atlantic in Europe, in North America from New Jersey to Newfoundland, and on the Pacific in Hawaii and Japan. Shimmering in the sun, the dense foliage usually remains submerged except when the tide pulls way back to reveal 6-to-12-inch growths clinging to coastal rocks. Fronds begin with narrow flattened stems that divide and subdivide in broad-forked fans ranging from reddish purple to reddish green.

Irish moss has many medicinal uses. The nineteenth edition of the *U.S. Dispensatory* states that *Chondrus crispus* is "particularly recommended in chronic pectoral affections, scrofulous complaints, dysentery, diarrhea and disorders of the kidneys and bladder." The extract carrageenin is used in over-the-counter and prescription medications for peptic and duodenal ulcers; it also yields an anticoagulant with a built-in time-release factor. Calcium chloride, a mineral compound found abundantly in Irish moss, acts as a heart tonic and maintains glandular balance.

In cooking, Irish moss is used primarily as a thickener (in stews, gravies, salads, aspics, pies, and puddings). Because of its high sulphur content, it tends to have a stronger sea odor than the other seaweeds; it is also tougher. Thus, the more easily manageable agar is often preferable; however, the two are interchangeable. It takes two times as much Irish moss as agar to gel liquid. Thus, where an agar recipe calls for 1 bar of kanten (1/4 ounce), 1/2 ounce Irish moss will suffice. A half-cup of the chopped dried seaweed equals approximately 1/2 ounce; this amount will gel 4 cups of water or thin liquid or 3 cups heavy liquid, soy milk, or liquid, with purees, cut fruits or cut vegetables added.

Before cooking with strongly flavored foods, the sea moss need only be rinsed once or twice and soaked in water for 10 minutes. For delicately flavored and sweet dishes, an additional rinse and 10-minute soaking is desirable.

Raw Roots Salad Serves 4–6

 1/3 cup dried Irish moss, soaked in cold water for 2 minutes, cleaned, and rinsed
 3 cups water
 1 bay leaf
 1/2 teaspoon each, mustard seed, caraway seed, dill weed
 1/4 teaspoon red pepper flakes
 3–4 peppercorns
 A pinch of allspice
 1/2 teaspoon honey
 1 tablespoon lemon juice
 1 beetroot, scrubbed clean and grated
 1 carrot, scrubbed clean and grated
 1 large Jerusalem artichoke (or several small), scrubbed and grated

Place the Irish moss in a saucepan with the water and spices, and cook over medium-high heat for 20 minutes. Strain through cheesecloth into a mixing bowl. Now stir in honey, lemon juice, and the grated roots, and transfer to a mold that has been rinsed in cold water (or lightly oiled) to prevent sticking. Allow to set. Serve as is or as part of a mixed salad.

Irish Moss Tomato Aspic Serves 4

This dish is extra-nutritious and delicious in winter, when fresh local tomatoes are not available.

 1/4 cup dried Irish moss, soaked in cold water, cleaned, and rinsed
 1 bay leaf
 3 allspice berries
 5 peppercorns
 1/2 teaspoon each, tarragon and paprika
 A handful of celery leaves, chopped
 2 cups tomato juice
 1 teaspoon olive oil
 1 teaspoon onion juice
 1–2 teaspoons lemon juice

Place Irish moss, spices, and celery leaves in a moistened muslin bag or square of cheesecloth and suspend in a saucepan. Add tomato juice and cook over medium heat, stirring frequently, for 20 minutes; press the bag with the back of a large wooden spoon to express all the gel. Remove pan from heat and bag from pan. Stir in olive oil and onion juice. Taste before adding lemon juice. Pour into individual molds lightly oiled with olive oil; chill until set. Unmold onto a bed of greens, or cut into wedges and toss with salad.

Cranberry Dessert Sauce Makes over 1 quart

This tart, but sweet and spicy sauce can be used to spark up cakes and even shortbread; or use it over ice cream, tofu cheese cakes, crepes, and/or puddings.

    1/2 cup dried Irish moss
    A 2-inch-long cinnamon stick
    A slice of fresh ginger root
    1 teaspoon coriander seed, bruised in a mortar
    1/4 teaspoon whole cloves, bruised in a mortar
    1 large cardamom pod, bruised in a mortar
    3 cups water
    1 pound new cranberries (about 4 cups), cleaned, and rinsed
    3 tablespoons honey
    1 cup cider or cranberry juice
    1-1/2 teaspoons freshly grated orange peel
    3 tablespoons currants
    1/4 cup slivered almonds

Combine Irish moss and the next five ingredients in a muslin bag or square of cheesecloth, and suspend in a large saucepan. Pour in the water and cranberries, and cook over medium-high heat until cranberries burst open and many have disintegrated (at least 20 minutes). Stir frequently as the sauce cooks, and press the bag against the side of the pan to express gel and flavorings. Now remove pan from heat and bag from pan. Stir in remaining ingredients, and set aside while the mixture thickens. Serve at room temperature, or chilled.

Variation: Double the amount of Irish moss called for and increase honey to 4 tablespoons. Mold for Cranberry Jelly, or use in place of agar fruit pie fillings.

Wine Jelly Makes 4 cups

    3-1/2 cups water
    1/2 ounce or 1/2 cup Irish moss, soaked in water for 10 minutes, rinsed, and drained
    Juice of 1 lemon
    3–4 tablespoons honey
    1/2 cup white wine or sake
    1/4 teaspoon cinnamon powder

Heat the water almost to a boil. Add Irish moss, turn down heat and simmer until it dissolves, about 15 minutes. Stir in lemon juice, honey, wine and cinnamon. Strain into jelly glasses or a mold.

Dulse

The Red Mittens

At the turn of the century, it was common to see tangy dried dulse hawked by vendors in New England railway stations and in the streets, much like pretzels. It was commonly served in bars as a snack, because its tasty saltiness was conducive to a lively emptying of beer mugs. Children were pacified with the chewy dried fronds, and it served adults as an inexpensive substitute for chewing tobacco.

Ten centuries ago, gathering dulse was a large industry in the western European islands. Icelandic sagas mention dulse as early as 961, and a Gaelic poem from the sixth century relates that the monks of St. Columba gathered it to feed the poor. Over the centuries the commercial exploitation of dulse dwindled until it was used only for winter animal fodder in Iceland, Scotland and Ireland; with the recent interest in natural foods, however, the industry is experiencing a revival. Today, dulse is gathered mostly in North America, from Newfoundland to Maine.

Most commercial dulse is gathered around the Bay of Fundy in Canada, in Nova Scotia, New Brunswick and Prince Edward Island. It is collected by hand at low tides, during the months from May to August, dried, and packaged. Dulse is rarely processed to any extent.

Rhodymenia palmata in Latin (and "dillisk," "red call," "crannough," or "Neptune's girdle" in various British Isles dialects), dulse is a typical red seaweed. It commonly grows on other seaweeds, acting as an epiphyte. It grows in distinctive flat, smooth, elastic fronds shaped like mittens (hence its Latin diminutive). Its stipes are short and round. The entire plant may measure 6–12 inches long and up to 6 inches wide. Dulse prefers strong currents in temperate zones; it reddens rocks and lodges below the low-water mark along the eastern and western Atlantic coasts and the northern Pacific coast of the United States.

Dulse has the highest concentration of iron of any food source, important in fortifying the blood. In addition, it is rich in potassium (vital to body fluid balance, and adrenal, kidney, and muscle function), and magnesium (needed daily to activate many enzyme systems, to aid in the synthesis of amino acids, to strengthen the muscles, and to sustain DNA and RNA production). Because of its high mineral content, dulse has been applied medically for a wide range of ailments. In 1695, in England,

Richard Martin found it to be efficacious in the prevention of scurvy. It was also used in the British Isles to induce sweating during fevers, and Icelanders used it as a remedy for seasickness. Its most spectacular application, however, is its recent use against Herpes virus (see *Time,* June 21, 1976).

Dulse lends itself to a wide variety of preparations, each one more delicious than the last. It has often appeared on New England dinner tables, fresh from the seas, as a salad vegetable or toasted. Once sun-dried, it may be freshened in water, drained, and treated like spinach. (One must look for occasional bits of shell that may still be clinging to this natural product!) The coastal residents of both the Old and New Worlds have commonly shredded it, fried it in butter, and seasoned it; it can be eaten fresh from the frying pan, used as a garnish for other vegetables, or blended in with mashed potatoes.

Alaskans gather dulse and fresh herring roe during the spawning season in summer, sauté them together lightly, and spike them with pepper for a succulent gourmet dish. Dulse often serves as a piquant alternative to bacon in European dishes, and as a tasty thickener for sauces and gravies. Freshened, dried dulse fronds can also be layered in casseroles or mixed into sandwiches.

The simplest or the most complex dishes—from relishes and pickles to salad dressings, shepherd's pies, baked beans, and even desserts—gain a special savoriness with the addition of dulse.

Dulse Simmered in Broth   Serves 1

> 1-1/2 cups homemade vegetable broth
> 1/4 cup dried dulse, torn into bite-size pieces
> A pinch of cayenne
> A sprinkling of minced herbs (fresh, if possible)

Gently simmer the dulse in the broth for 5 minutes. Serve seasoned with cayenne and herbs.

Nut Milk Soup with Dulse   Serves 4

Milk:

> 1 cup raw cashews
> 4 cups water

1 tablespoon olive oil
1/4 cup chopped shallots, leek, or any member of the garlic-onion family
1 teaspoon tarragon
1/2 teaspoon kelp powder
A pinch of rosemary
1-1/2 cups dried dulse, torn into bite-size pieces
A dash of cayenne or freshly ground black pepper
A few sprigs of parsley

Prepare "milk" in blender. In a soup pot, heat the olive oil, and sauté the next four ingredients for about 5 minutes. Add milk, then dulse; purée in blender until smooth. Serve with a sprinkling of pepper and garnished with parsley sprigs.

Cream of Dulse & Potato Soup Serves 4

  1/2 cup dried dulse, immersed in cold water to cover for
      1 minute; cleaned, drained, and set aside
  1 tablespoon olive oil
  1 pound potatoes (mealy Idaho type), cubed
  White portion of leek, chopped
  1/2 teaspoon each rosemary powder, sage, and thyme
  6 cups homemade vegetable soup stock
  A dash of red or black pepper
  1/4 cup minced fresh parsley

Heat oil in a soup pot, and sauté potatoes, leek, and herbs until leek is transparent. Pour in stock, cover, and cook until potatoes are tender (about 10 or 15 minutes). Add dulse and purée mixture in a blender; reheat until heated through. Serve dusted with pepper and garnished with fresh parsley.

Dulse Chowder Serves 6

  1 tablespoon safflower or corn oil
  2 stalks celery (including tops), diced or sliced diagonally
  1 onion, chopped
  1 clove garlic, slivered
  2/3 cup diced carrot, potato, or zucchini
  1/2 teaspoon each oregano, sage, thyme
  1/4 teaspoon each cumin powder and paprika
  1 cup corn (freshly shucked or home-canned)

Soup base:

  2 cups water
  2 tablespoons arrowroot starch
  1/2 teaspoon kelp powder
  1/2 cup raw cashews
  1 cup corn
  1/2 cup celery leaves

  4 cups homemade vegetable broth or water
  1-1/2 cups dried dulse, torn into bite-size pieces

Heat oil in a large soup pot, then sauté the next six ingredients together until onion is transparent. Add corn and cook for 5 minutes. Meanwhile, purée soup base in a blender, then stir the purée into the pot. Add broth or water, and dulse. Reduce heat and simmer for 5 minutes before serving.

Dulse Dressing
Makes 1/2 cup

Serve with strong greens: escarole, chicory, spinach, etc.

> 1/2 cup dried dulse, freshened in water and drained
> 4 umeboshi plums, finely chopped
> 1 scallion chopped
> 1/2 cup oil (half peanut and half safflower)

Chop dulse fine. Place all ingredients in blender and blend well.

Mediterranean Salad with Dulse
Serves 4

> 3-4 citrus fruits, peeled and roughly chopped
> 1 cup dried dulse, rinsed, cleaned, and drained

Dressing:

> 1/3 cup safflower oil
> 2 tablespoons olive oil
> 2 tablespoons red wine vinegar
> 1-2 tablespoons lemon juice
> 1/2 teaspoon each oregano and cumin powder
> 1/4 teaspoon kelp powder
> Freshly ground black pepper, to taste

> A bed of salad greens
> Rings of raw purple onion

Toss fruits and dulse with dressing; marinate for 15 minutes. Arrange on salad greens and top with onion rings.

Dulse & Potato Salad
Serves 4

Dressing:

> 2 tablespoons safflower oil
> 2 tablespoons sunflower seed butter (or 2 tablespoons sunflower seeds, pounded with pestle)
> 2 tablespoons apple cider vinegar
> 1/2 tablespoon lemon juice, approximately
> 1/2 cake tofu, mashed thoroughly
> 1 teaspoon basil
> 1/4 teaspoon paprika

> 8 large lettuce leaves, of a sweet variety
> 1-2 handfuls of mixed bitter greens, chopped or torn
> 1 cup dried dulse fronds, freshened in water, rinsed, and drained
> 3/4 pound new potatoes, cooked whole and cooled enough to handle
> 1 red bell pepper, cut into large bite-size slivers
> Freshly ground black pepper, to taste

Prepare dressing by beating ingredients together well; set aside. Tear lettuce leaves into the bottom of a large bowl and toss with greens and dulse. Top with potatoes and red peppers. Serve dusted with black pepper, accompanied by dressing.

Sautéed Dulse
Serves 4

    2 cups dried dulse, torn into bite-size pieces
    2 tablespoons vegetable oil
    A sprinkling of cayenne or freshly ground black pepper

Cover dulse with cold water and allow to soak for 1 minute; clean dulse and drain. Heal oil in enamel skillet, and sauté dulse for 3 to 5 minutes, stirring from time to time. Season to taste with pepper.

Greek-Style Potato Spread
(Skordalia)
Serves 4

This spread is delicious with crackers or chips or in sandwiches or pita bread.

    6 cloves garlic
    4 medium potatoes, boiled and mashed
    1/2–3/4 cup bread crumbs (preferably dark or Italian)
    1/2 cup dried dulse, freshened in water, drained, and cut into bite-size pieces
    1/2 cup oil (half olive and half safflower)
    1/3 cup vinegar (half rice and half cider)
    1/3 teaspoon kelp powder

Mash garlic and add to potatoes, along with bread crumbs and dulse, stirring well. Now add oil, a few tablespoons at a time, alternating with a few tablespoons of vinegar; beat thoroughly. Add kelp powder to taste, and beat until thick and smooth. Cover and refrigerate for several hours. Use within a week.

Variation: Substitute cooked chickpeas (drained) for the potatoes and/or add 1/2 cup chopped nuts or sunflower seeds.

Greens & Sea Greens
Serves 4

Dulse's striking, iron-rich taste complements the mellow taste of many head lettuces.

    8 large lettuce leaves, torn into pieces
    1 cup dried dulse fronds, freshened in water and drained
    3 ounces mushrooms, sliced
    3 scallions, cut diagonally into 1-inch lengths

Toss all ingredients with dressing; (for a lemon dressing see Wakame Vinaigrette).

Dulse & Sprouts Sandwich
Serves 1

Dressing:

    1 teaspoon lemon juice
    4 tablespoons tahini

    2 large slices whole grain bread or 1 whole wheat pita
    1/2 cup alfalfa or lentil sprouts
    1/2 cup dried dulse, freshened in water, drained and roughly chopped
    1/2 cake tofu, thinly sliced (optional)
    3–4 large lettuce leaves

Prepare the dressing, and use as a spread on the bread. Fill sandwich with remaining ingredients.

Savory Vegetable Pie with Dulse

Serves 6

This peasant dish is easy to fix and, when served with a salad, is a welcome one-dish supper.

1 deep-dish pie crust, 9-1/2 inches
2 tablespoons vegetable oil
1 clove garlic, minced
1 small onion, chopped
1/4 teaspoon caraway seeds
1/2 teaspoon marjoram
1/4 teaspoon sage
1/4 teaspoon kelp powder

Vegetable mixture:

1 small bell pepper, cut into small strips
1/2 pound zucchini, sliced into thin half-moons

or

1 stalk celery, cut crosswise into thin slices
1/2 pound trimmed broccoli, cut into bite-size pieces

1-1/2 cups dried dulse, torn into bite-size pieces
A handful of minced fresh parsley
Freshly ground black pepper, to taste
3-1/2 cups cooked, puréed kidney or other red beans, hot or at room temperature
Paprika, to taste

Preheat oven to 350°. In a heavy pan coated with oil, sauté garlic, onion, and spices for 2 minutes. Add vegetable mixture and sauté for 5 minutes. Stir in dulse, parsley, and pepper, and sauté for 1 minute more.

Partially fill crust with 1 cup bean purée, top with a sprinkling of paprika, and cover with a layer consisting of one-half of vegetable mixture. Repeat until all ingredients are used, topping with 1-1/2 cups purée. Sprinkle with paprika and bake for approximately 25 minutes, until crust turns golden. Allow to set and cool for 5 minutes before slicing.

North African Casserole (*Tagine*) — Serves 4

3 large stalks Swiss chard, chopped
1 medium onion, chopped
1/2 cup chopped parsley (preferably broad-leaf)
1/2 cup vegetable or olive oil
1/2 teaspoon cayenne
1/2 teaspoon kelp powder
1/4 teaspoon each nutmeg and ground coriander
1/4 cup water
1/2 cup dried dulse, freshened in water, drained, and chopped
1/2 cup barley (or rice or lentils)

Combine the first eight ingredients in a casserole, and simmer, covered, for 30 minutes. Stir in dulse and barley, and continue simmering for about 30 minutes, or until the barley is tender.

Mashed Potatoes with Dulse — Serves 4

1 pound new potatoes, cooked and peeled
1 tablespoon olive oil
1 clove garlic, pressed or minced fine
1/2 cup dried dulse, torn into pieces
Oil for frying
Black or red pepper, to taste

Mash potatoes with oil and garlic, beating in as much cooking liquid as necessary to reach desired consistency. Quickly stir-fry dulse in oil, then beat dulse in with potatoes. Season with pepper.

Variation: Use mixture to form Sea Vegetable Croquettes.

Dulse Hash — Serves 4

1 pound potatoes
2 tablespoons vegetable oil
1 teaspoon black mustard, poppy, or chia seeds
2 cloves garlic
1/3 cup chopped mushrooms
1/3 cup chopped onions
1/3 cup chopped sweet pepper
1 cup dried dulse, torn into bite-size pieces
A sprinkling of cayenne and paprika

Boil potatoes until partially cooked (so that they can be pierced by a knife but not a fork). Allow to cool, then peel, if desired, and cube. Heat oil in a large heavy skillet. Add seeds and potatoes, frying until potatoes are pale gold; remove potatoes from pan. Reheat remaining oil and stir-fry garlic, mushrooms, onions, and pepper for 5 minutes. Stir in dulse and potatoes, and continue cooking, stirring constantly, for 5 to 10 minutes more. Sprinkle with cayenne and paprika just before serving.

Dulse and Kale
Serves 4

- 4 tablespoons olive oil
- 2 cloves garlic, chopped
- 2 pounds (or 2 quarts) kale, roughly chopped
- 1/2 cup dried dulse, freshened in water, drained, and roughly shredded
- 1/3–1/2 cup pignoli nuts (or cashew pieces)

Heat olive oil in a large heavy skillet. Sauté garlic for 30 seconds, then add the kale and sauté for 2 or 3 minutes. As soon as the kale begins to deepen in color, add dulse and nuts, mixing well. Cover and let steam until the oil has thoroughly penetrated (about 2 minutes). Do not overcook. Serve immediately, as an accompaniment to fried dishes.

Variation: Substitute any strong green, such as mustard greens, collard greens, or spinach for the kale.

Okra Dulse Deluxe
Serves 4

- 2 tablespoons oil
- 1 clove garlic, chopped
- 1/2 pound okra
- 1/4 pound fresh mushrooms, cut into halves
- 1 cup dried dulse, freshened in water, drained and chopped
- 1 (8-ounce) can water chestnuts, cut lengthwise into halves
- 2 tablespoons natural soy sauce (shoyu)
- 1 tablespoon lemon juice
- 1/2 teaspoon chili powder
- A sprinkling of cayenne (optional)

Heat the oil in a skillet or wok and sauté the garlic for 30 seconds or so. Trim okra ends, if desired. Add okra and next three ingredients to garlic, and sauté for 1 to 2 minutes more, until the okra and mushrooms begin to darken in color. (Do not overcook: the vegetables should retain their crispness.) Stir in soy sauce and lemon juice, and turn off heat. Season with chili powder and, if used, cayenne.

Dulse and Zucchini Cornucopia
Serves 2

- 1 tablespoon vegetable oil
- 2 scallions, cut into 1-inch lengths
- 1/2 teaspoon basil
- 1 teaspoon thyme
- 2 ounces mushrooms, sliced
- 1 small red bell pepper, slivered
- 6–8 ounces zucchini, cut into thin slices or matchsticks
- 1 cake tofu, cubed (optional)
- 1/2 cup fenugreek seed sprouts (optional)
- A sprinkling of paprika
- 2 cups chopped spinach
- 1/2 cup chopped dandelion leaves (optional)
- 1 cup dried dulse fronds, rinsed in fresh water just prior to cooking
- Dash of water

Heat oil in a wok or skillet. Add the next six ingredients and stir-fry for 3 to 5 minutes, then reduce heat to low. Now add tofu cubes, if used, and the next five ingredients, mixing lightly. Splash with water and quickly cover. Steam for 2 minutes, or until greens are wilted and tofu is heated through.

Sea Vegetable Croquettes

Makes 4 croquettes

Traditional recipes for Great Britain's Bannock Cakes vary: fresh sea vegetables may be mixed with oatmeal and fried in lard; or they may be beaten and dried, then added to porridge or fried into "cakes." Thick rice and corn porridge are suitable. Croquettes may also be made from mashed potatoes, peas, or beans, or puréed vegetables such as squash, pumpkin, or carrots.

 1 cup rolled oats
 2 tablespoons dulse flakes
 2 cups vegetable purée
 2–3 tablespoons vegetable oil

Mix oats and dulse. Add enough to purée to thicken to desired consistency. Form croquettes and coat with more oats and dulse. After chilling for at least 15 minutes, sauté croquettes in vegetable oil for 5 to 7 minutes on each side, until heated through and crisp; or bake in a moderate oven for 20 to 25 minutes.

Pickled Celery and Dulse

Makes 2–3 cups

Marinade:

 1/2 cup rice vinegar
 1/2 cup olive oil (or a combination of half virgin olive, half safflower)
 1 tablespoon sake
 2 bay leaves
 1 teaspoon basil
 1/2 teaspoon kelp powder

 3–4 stalks celery (including leaves), cut into bite-size pieces
 3–4 cups boiling water
 2/3 cup dried dulse, freshened in water and drained

Prepare marinade and set aside. Drop celery into the boiling water, then turn off heat immediately; cover and let stand for 3 or 4 minutes. Now drain celery and combine with the dulse in a sterilized jar with a tight-fitting lid; pour in marinade and seal. Refrigerate for at least 12 hours. Refrigerated, this pickle should keep for several weeks.

Party Favors
Serves 4

Filling:

    3 tablespoons tahini
    1/2 tablespoon mustard powder
    1 tablespoon hot water, approximately
    1/2 tablespoon natural soy sauce (shoyu)
    A pinch of kelp powder

1 cup dried dulse, freshened in water, wrung dry, and broken into 1-inch lengths
3 sheets of nori, cut into eighths

Prepare the filling and mix the dulse in well, thinning with hot water as needed, to make a creamy paste. Wrap about a teaspoonful of filling in each piece of nori, and secure with a foodpick.

Variation: You may also wish to wrap rice with the filling in the middle. (See Nori-maki.)

Dulse Pudding
Serves 4–6

    3 cups soy or nut milk
    3 tablespoons arrowroot starch
    1/4 teaspoon sea salt
    1 inch-long cinnamon stick
    3/4 cup millet
    4 tablespoons honey
    2 cups dried dulse fronds
    1 teaspoon grated orange rind
    1 teaspoon ground coriander
    2 apples or oranges, peeled, seeded, and chopped (optional)

Over medium-high heat blend the first three ingredients in the top of a double boiler. Add cinnamon stick and millet, and cook, stirring frequently, for 5 minutes. Reduce heat, stir in remaining ingredients, and simmer, without stirring, until thick (about 45 minutes).

Sources of Sea Vegetables

The Oriental sea vegetables—nori, kombu, wakame, agar, arame, and hijiki—can be found in their dried forms in most Japanese, Korean, and Chinese markets. Together with dulse and Irish moss, they are also available in the health food stores that have sprung up in nearly every city across the United States. Dulse is gathered from our own New England coast and Canada, and distributed by Atlantic Mariculture; Irish moss is available through National Health Foods, Inc. Many herb suppliers offer dried kelp and dulse, agar, Irish moss, and bladderwrack, a brown kelp from which tea is made. A number of guides to the uses of wild plants include sections on seaweeds, covering some local varieties not now commercially produced. To discover local outlets and cooking ideas, try a nearby Japanese restaurant, ask in the Chinese area, or urge the local health food store to stock more varieties of sea vegetables.

In addition to the dried sea vegetables, you may wish to try powdered kombu tea or powdered kombu and umeboshi tea; seaweed crackers (*senbei*), which come in many varieties; kombu candy; and various Chinese and Japanese seaweed pickles. These, too, are stocked in health food stores and Oriental markets.

Major Distributors

Atlantic Mariculture, Ltd.
P.O. Box 2368
Dartmouth, Nova Scotia B2W 3EO

Chico San Company
Chico, CA 95926

Erewhon Trading Company
342 Newbury St.
Boston MA 02115

Erewhon
8001 Beverly Blvd.
Los Angeles CA 90048

Japan Food Corporation
P.O. Box 6096
Long Island City NY 11106

Japan Food Corporation
(Hime brand)
San Francisco CA 94080

Nichols Garden Nursery
1190 North Pacific Highway
Albany OR 97321

Nishimoto Trading Company
(Shirakiku brand)
139 Ninth St.
Jersey City NJ

The Shepards
Franklin ME 04634

Soken Trading Company
Sausalito CA 94965

Westbrae Natural Foods
4240 Hollis St.
Emeryville, CA 94608

Nutritional Value of Sea Vegetables

Name	Nutrients (%)				Minerals (mg/100g)						
	Fiber	Prot.	Fat	Carbo.	Ca.	I	Fe	Mg	Ph	K	Na
Agar			0.3	16.3	567	0.2	6.3		22		
Dulse	1.2	20.—30.	3.2	44.2	296	8	150	220	267	8060	2100
Hijiki		5.6	0.8	29.8	1400		30		56		
Irish Moss			3.2		885		8.9		157	2844	2900
Kelp	3.0		1.1—1.8	51.9	1093	150	100	760	240	5273	3000
Kombu		7.5	1.1	51.9	800	76.2			150		
Nori-Green		34.2	0.6	40.5	470		23		580		
-Red		34.2	0.7	40.5	470		23		580		
Wakame	3.6	12.7	1.5	47.8	1300	7.9			260		1100
Spinach	0.6	3.2	0.3	4.3	93	0.036	3.1	88	51	470	

Name	Vitamins (mg/100g)					
	A*	B_1	B_2	B_5	B_{12}	C
Agar						
Dulse		.63	.50	1.69		24—49
Hijiki	555	.01	.02	4.0		
Irish Moss						
Kelp		.09	.33	5.7	0.5—1.0	13
Kombu	430	.08	.32	1.8	0.5—1.0	11
Nori-Green	960	.06	.03	8.0	0.7	10
-Red	6000—11000	.12—.25	.20—1.24	2.0—10.0	0.7	20
Wakame	140	.01	.02	10.0	0.6	15
Spinach	8100	.10	.20	0.6		51

*Vitamin A is given in international units.

Nutrition Chart Sources

Krochmal, Connie and Arnold. *A Naturalist's Guide to Cooking with Wild Plants*. New York: The New York Times Book Company, 1974.

Food Heritage Health Research. *Composition and Facts About Foods*. California, 1968.

Shufu no Tomo Sha. *Kaiso Ryori*. 1975.

Muramoto Naboru. *Healing Ourselves*. New York: Hearst Publishing.

U.S. Department of Agriculture. *Composition of Foods*. Agriculture Handbook No. 8, 1975.

Bibliography

Ash, A.S.F. "Seaweed as Food." *Food Preservation Quarterly,* December, 1954.

Black, W.A.P. *Seaweeds and Their Value in Foodstuffs.* Nutrition Society Proceedings, 1953.

Boney, A.D. "Aspects of the Marine Biology of the Seaweeds of Economic Importance." *Advances in Marine Biology,* no. 3 (1965).

Bosco, Dominick. "Kelp and Other Gifts from Neptune's Garden." *Prevention,* October, 1975.

Brooklyn Botanic Garden, *Japanese Herbs and Their Uses.* Handbook #57 (1976).

Chapman, Esther. *How to Use the 12 Tissue Salts.* New York: Pyramid Books, 1976.

Chapman, Valentine J. *The Algae.* New York: St. Martin's Press, 1962.

———*Seaweeds and Their Uses.* New York: Methuen & Co., Ltd., 1950.

Cheng, T. "Production of Kelp—A Major Aspect of China's Exploitation of the Sea." *Economic Botany,* no. 3 (1969).

Coker, R.E. *This Great and Wide Sea: An Introduction to Oceanography and Marine Biology.* New York: Harper and Row, 1962.

Collier, A. "The Significance of Organic Compounds in Sea Water." *Transcript of the 16th American Wildlife Conference,* 18 (1953).

Cowen, Robert C. *Frontiers of the Sea.* New York: Doubleday, 1969.

Dawson, E. Yale. *How to Know the Seaweeds.* Dubuque, Iowa: William C. Brown, 1956.

DeGuerin, Basil C. "Carrageen in the Channel Isles." *Food Manufacture,* March, 1946.

Dixon, Peter F. *Biology of the Rhodophyta.* New York: Hafner Press, 1973.

Duddington, C.C. *Flora of the Sea.* New York: Thomas Crowell, 1967.

Food and Agriculture Organization of the United Nations. *The Provision of More Adequate Supplies of Edible Protein.* F.A.O. Report, 60451-65-WM, 1965.

Fowden, L. "A Comparison of the Composition of Some Algal Proteins." *Annals of Botany,* July, 1954.

Fryer, Lee, and Simmons, Dick. *Food Power from the Sea—The Seaweed Story.* New York: Mason/Charter, 1977.

Gaul, Albro. *The Wonderful World of the Seashore.* New York: Appleton-Century Crofts, 1955.

Gibbons, Euell. *Stalking the Good Life.* New York: David McKay Co., 1966.

Greulach, Victor A., and Adams, J. Edison. *Plants, An Introduction to Modern Botany.* New York: John Wiley & Sons, 1962.

Grièves, M. *A Modern Herbal.* 2 vols., 1931. Reprint. New York: Dover Publications, 1971.

Guberlot, Muriel L. *Seaweeds at Ebb Tide.* Seattle: University of Washington, 1956.

Guiseley, Kenneth B. *Seaweed Colloids.* New York: John Wiley & Sons, 1968. Reprinted from Kirk-Othmer, *Encylopedia of Chemical Technology.*

Hall, John Betal. Blue-Green Algae: Current Research. Vol.IV (1974).

Hatch, M.T.; Ehresmann, D.W.; Deig, E.F.; Vedios, N.A. *Further Studies on the Chemical and an Initial In Vivo Evaluation of Antiviral Material in Extracts of Macroscopic Marine Algae.* Berkeley, California: Naval Biosciences Laboratory, University of California.

Hesp, R., and Ramsbottom, B. "The Effect of Sodium Alginate in Inhibiting Uptake of Radiostrontium by the Human Body." *Nature,* December 25, 1965.

Hewitt, James. *About Sea Foods.* Sussex, England: Lewis Press Wightman and Co., Ltd., 1964.

Hillson, C.J. *Seaweeds, A Color-Coded Illustrated Guide to Common Marine Plants of the Eastern Coast of the United States.* Philadelphia: University of Pennsylvania, 1977.

Hunter, Kathleen. *Health Foods and Herbs.* New York: Arco Publishing Co., 1971.

Jackson, Daniel F. *Algae and Man: Proceedings.* Louisville Ky.: NATO Advanced Study Institute, 1962, 1964.

Jarvis, D.C. *Folk Medicine: A Vermont Doctor's Guide to Good Health.* New York: Holt, Rinehart & Winston, 1958.

Kavaler, Lucy. *The Wonders of Algae.* New York: John Day Co., 1961.

Kingsbury, John M. *Seaweeds of Cape Cod and the Islands.* Old Greenwich, Conn.: The Chatham Press, Inc., 1969.

Kreig, Margaret B. *Green Medicine: The Search for Plants that Heal.* Skokie, Ill.: Rand McNally, 1964.

Krochmal, Connie and Arnold. *A Naturalist's Guide to Cooking with Wild Plants.* New York: Quadrangle/The New York Times Book Co., 1975.

Kurogi, M. "Recent Laver Cultivation in Japan." *Fishing News International,* July–September, 1963.

LaPointe, B.E., et al. *The Mass Outdoor Culture of Microscopic Marine Algae.* Contribution #3609, Woods Hole Oceanographic Institute #47, Harbor Branch Foundation.

Lavey, John, and Tischer, Robert G. *Food from Algae: A Review of the Literature.* Quartermaster Food and Container Institute for the Armed Forces Library Bulletin, 1958.

Levring T.; Hope, H.A.; and Schmid, O.J. "Marine Algae: A Survey of Research and Utilization." *Botany Marine Handbook,* 1969.

Lewin, Ralph A. *The Physiology and Biochemistry of Algae.* New York: Academic Press, 1962.

Lucas, Richard. *Common and Uncommon Uses of Herbs for Healthful Living.* 1969. Reprint. New York: Arc Printing, 1972.

Madlener, Judith Cooper. *The Sea Vegetable Book.* New York: Clarkson N. Potter Inc., 1977.

Mautner, Henry. "The Chemistry of Brown Algae." *Economic Botany,* April–June, 1954, pp. 174–192.

Muramoto, Naboru. *Healing Ourselves.* New York: Avon, 1973.

Naylor, J. *Production, Trade and Utilization of Seaweeds and Seaweed Products.* Rome: Food and Agriculture Organization of the United Nations, 1976.

Newton, Lily. *Seaweed Utilization.* London: S. Low, 1951.

———"Uses of Seaweed." *Vistas in Botany,* 1963.

Nova Scotia Research Foundation. *Selected Bibliography in Algae.* Dartmouth, Nova Scotia.

Oishi, Keiichi, and Takeo, Harada. *The Book of Kombu* (Kombu no Hon). Japan: Kanki Press, 1977.

Okazaki, Akio. *Seaweeds and Their Uses in Japan.* Tokyo: Tokai University Press, 1971.

Palmer, E.L. "The Marine Algae—In the Sea's Weeds May Lie the Future's Insurance Against Starvation." *Natural History,* 1961.

Phycological Society of America. "Abstracts of papers scheduled for the combined meetings of the International Seaweed Symposium, the Phycological Society of America, and the International Phycological Society at the University of California, Santa Barbara, California, August 20–27, 1977." *Journal of Phycology,* June, 1977.

Powell, Eric F.W. *Kelp, the Health Giver.* Northants, England: Health Sciences Press, 1968.

Prescott, Gerald W. *The Algae: A Review.* Boston: Houghton-Mifflin, 1968.

Prevention staff. "Algin for Heartburn." *Prevention,* August, 1975.

Proceedings of the First International Seaweed Symposium. Edinburgh, 1953.

Proceedings of the Second International Seaweed Symposium. Trondheim et al, editors. New York: Pergamon Press, 1956.

Proceedings of the Third International Seaweed Symposium. Galway, 1958.

Proceedings of the Fourth International Seaweed Symposium. Biarritz, A.D.; de Virville, Doug; and Feldman, J., editors. New York: Pergamon Press, 1961.

Proceedings of the Fifth International Seaweed Symposium. Halifax, editor. New York: Pergamon Press, 1966.

Proceedings of the Sixth International Seaweed Symposium. Margalet, R., editor. Madrid, 1968.

Proceedings of the Seventh International Seaweed Symposium. Trondheim, editor. New York: John Wiley and Sons, 1974.

Proceedings of International Symposium at Syracuse University. *Algae, Man and the Environment.* Daniel F. Jackson, editor. Syracuse. N.Y.: Syracuse University Press, 1968.

Reed, Minnie. *The Economic Seaweeds of Hawaii and Their Food Value.* Washington, D.C.: Washington Printing Office, 1907.

Rodale, J.I. *The Complete Book of Minerals for Health.* Emmaus, Pennsylvania: Rodale Press, 1972.

———*The Complete Book of Vitamins for Health.* Emmaus, Pennsylvania: Rodale Press, 1969.

Round. F.E. *Biology of the Algae.* 2nd ed., New York: St. Martin's Press, 1974.

Sackheim, George I., and Schultz, Ronald M. *Chemistry for the Health Sciences.* New York: Macmillan, 1973.

Salisbury, O.M. *The Customs and Legends of the Tlingit Indians of Alaska.* New York: Bonanza, 1952.

Sanford, F.B. *Seaweeds and Their Uses.* Fisheries Leaflet #469, Washington, D.C.: U.S. Department of the Interior, 1958.

Schwimmer, Morton and David. *The Role of Algae and Plankton in Medicine.* New York: Grune & Stratton, 1955.

Schooner, Mary. "Red Algae from Seaweed to Kanten." *East-West Journal,* Vol. 1, no. 9 (1971).

Scientific Affairs Division, NATO Advanced Study Institute, (sponsor). *Algae and Man.* 1962.

Shufu no Tomo Publishers. *Seaweed Cooking* (Kaiso Ryori Hachijusshu). Tokyo, Japan: Shufu no Tomo Publishers, 1975.

Shurtleff, William, and Aoyagi, Akiko. *The Book of Tofu.* Brookline, Ma.: Autumn Press, 1975.

──────*The Book of Miso.* Brookline, Ma.: Autumn Press, 1976.

Smith, Donald, and Young, Gordon. "The Combined Amino Acids in Several Species of Marine Algae." *Journal of Biological Chemistry,* December 1955.

Stephenson, W.A. *Seaweed in Agriculture and Horticulture.* London: Faber and Faber, 1968.

Taylor, William R. *Marine Algae of the Northeastern Coast of North America.* Ann Arbor: University of Michigan Press, 1960.

──────*Marine Algae of the Eastern Tropical and Subtropical Coasts of the Americas.* Ann Arbor: University of Michigan Press, 1960.

Tiffany, Lewis H. *Algae: The Grass of Many Waters.* Springfield, Ill.: C.C. Thomas, 1968.

Triffitt, J.T. "The Binding of Calcium and Strontium by Alginates." *Nature* 217, 1968.

Taub, Harold J. "Kelp Can Guard Against Radiation Dangers." *Prevention,* August, 1972.

Time Magazine. "Succor from Seaweed: Herpes I and II." June 21, 1976.

Tseng, C.K. "Gloiopeltis and Other Economic Seaweeds of Amoy, China." *Lingnan Science Journal,* vol. 12 (1933).

──────"Seaweed Products and Their Uses in America." *Chemurgic Digest,* 1946.

──────"Seaweed Resources of North America and Their Utilization." *Economic Botany,* 1947.

Tsuchiya, Yasuhiko, and Suzuki, Y. "The Savor of a Laver Porphyra tenera." *Japanese Society of Scientific Fisheries Bulletin,* 1955.

Turrentine, J.W. *Potash from Kelp: Early Development and Growth of the Giant Kelp Macrocystis pyrifera,* Washington, D.C.: U.S. Government Printing Office, 1923.

Turner, N.C. *Notes on Haidu Indian Edible Seaweeds.* Victoria, British Columbia: Provincial Museum, 1974.

U.S. Department of Agriculture, Agricultural Research Service. *Composition of Foods.* Agriculture Handbook, No. 8 (1975).

Vogel, Virgil J. *American Indian Medicine.* Norman, Oklahoma: University of Oklahoma, 1970.

Water Resources Science Information Center, editor. *Algae Abstracts.* 2 vols. New York: Plenum Publishing Corp., vol. 1, to 1969, vol. 2, 1970–1972.

Weisz, Paul B., and Fuller, Melvin S. *The Science of Botany.* New York: McGraw-Hill, 1962.

Zagic, E.J. *Properties and Products of Algae.* New York: Plenum Publishing Corp., 1970.

Zanevid, J.S. *Economic Marine Algae of Tropical South and East Asia and Their Utilization.* Special Publication #3, Bangkok: Food and Agriculture Organization of the United Nations Regional Offices, 1955.

Zenekevitch, L. *Biology of the Seas of the U.S.S.R.* New York: Interscience Publishers, 1963.

Recipe Index

African peanut soup, 78
Agar
  Almonds with, 108
  Strands, 106
  Tofu, 104
Almond sweet, 108
Almonds, tangy, 30
Annin dofu, 108
Appetizers
  Agar tofu, 104
  Kombu chips, 53
  Kombu-wrapped tofu, 47
  Korean-style nori, 77
  Nori-wrapped fried tofu, 84
  Nori rolls, 85
  Nuts and nori, 78
  Open face sandwich, 66
  Party favors, 126
  Potato spread, 121
  Sweet potato and hijiki, 99
Apple salad, 93
Apples, used in
  Dulse pudding, 126
  Tapioca pudding, 89
Apple-agar topping, 113
Arame
  Cabbage and, 59
  Carrots and, 58
  Curried, 58
  Hot and sour soup, 55
  Sauteed, 56
  Scrambled, 56
  Stir-fried, 60
  Tamale pie with, 57
Arame and cabbage in mustard sauce, 59
Arame and carrots with hot peanut sauce, 58
Asparagus, 83
Aspic
  Bouillon, 103
  Molasses spice, 106
  Lemon bouillon, 103
  Citrus jam, 111
  Grape jello cubes, 109
  Grapefruit dessert, 109
  Japanese, 105
  Layered avocado, 104
  Mizu yokan, 112
  Molded fruit, 106
  Pear jello, 107

Pumpkin pie, 110
Spiced pears, 107
Strawberry jam, 111
Avocado
  Layered with vegetable mold, 104
  Split, 46

Baked spiced pears in cider aspic, 107
Baked stuffed pumpkin, 98
Bamboo shoots
  Mock crabmeat with, 100
  Wakame and, 73
  Wakame sea salad with, 66
  Wontons using, 79
Bamboo shoots & wakame salad, 66
Bannock Cakes, 30
Barley, 123
Beans and peas; Beans
  Azuki, 112
  Black, 92
  Kidney in shepherd's pie, 98
  Kidney and string beans, 60
  Kidney in vegetable pie, 122
  Lima, 72
  Lentil salads, 82
  Lentil spread, 74
  Lentils, in casserole, 123
  Snap, 60
  Soybeans, Japanese-style, 48
  Soybeans, wakame stew and, 69
  Soybean milk, 72
  String, with wakame vinaigrette, 67
Beans and peas; Peas
  Black-eyed, 48
  Chick, in Mediterranean salad, 46
  Chick, in potato spread, 121
  Snow, with creamy dressing, 73
  Snow, in soup, 65
  Snow, in mock crabmeat, 100
  Snow, with potato and hijiki, 99
  Wakame and bamboo shoots with, 73
Black-eyed peas with kombu, Texas-style, 48
Blueberry-agar topping, 113
Bouillon aspic, 103
Bouillon, kombu, used in
  Bouillon aspic, 103
  Clear soup, 62
  Wakame and chrysanthemum soup, 65
Braised sprouts & wakame, 74

Broccoli
  Salad, 56
  Vegetable pie using, 122
  Stew, 71
Broccoli & hijiki soup, 92
Broccoli salad with arame and mustard dressing, 56
Broccoli stems, 92
Brussel sprouts, 57
Brussel sprouts and arame teriyaki, 57
Butter, peanut, 78

Cabbage in cole slaw, 80
Candy, 52
Candied kombu, 52
Carrots
  Aspic using, 105
  Cole slaw using, 80
  Dulse chowder using, 119
  Hijiki with, 100
  Raw salad using, 115
  Sauce using, 58
  Stir-fried, 60
  Tsimmas using, 86
Cauliflower, in
  Wakame vinaigrette, 67
  Wok, 83
Celery
  Curried hijiki with, 93
  Dulse chowder with, 119
  Pickled, 125
  Vegetable pie using, 122
Chard, Swiss
  Casserole using, 123
  Miso soup using, 63
Chinese cucumber salad, 45
Chestnuts, water, 124
Chick-peas, *see beans and peas*
Citrus jam, 111
Clear soup with wakame, 62
Cole slaw, 80
Color show, 96
Condiments
  Citrus jam, 111
  Henry's favorite, 81
  Kombu chips, 53
  Kombu, pickled, 51
  Kombu, spicy, 50
  Kombu-wrapped tofu, 47
  Kombu *tsukudani*, 50

Nori, 87
Nori and cucumber, 80
Nori-miso "pickle", 89
Nori *tsukudani*, 85
Party favors, 126
"Pickle", nori-miso, 89
Pickles, garlic and kombu, 52
Pickled cucumbers and kombu, 52
Pickled kombu, 51
Spicy kombu, 50
Strawberry jam, 111
Sunchoke and kombu, 51
Sweet and spicy, 52
Toasted nori, 76
Umeboshi and nori, 89
Wakame, 72
Corn, used in
  Dulse chowder, 119
  Wakame succotash, 72
Cranberry dessert sauce, 116
Cream of dulse & potato soup, 119
Creamy avocado & kombu soup, 44
Crepes, Nori, 90
Croquettes, Sea Vegetable, 125
Crust (pie), *see pies, crust*
Cucumber
  Nori and radish with, 80
  Wakame and noodle salad with, 67
Cucumbers & kombu pickled in shoyu, 52
Cucumber, wakame, and clear noodle salad, 67
Curried Arame, 58
Curried hijiki and apple salad, 93

Daikon, see radish, daikon
Dashi, 44
Deep-fried kombu chips, 53
Dessert
  Almond sweet, 108
  Baked spiced pears, 107
  Candied kombu, 52
  Cranberry sauce, 116
  Dulse pudding, 126
  Grape jello cubes, 109
  Grapefruit, 109
  Mizu yokan, 112
  Molasses spice aspic, 106
  Molded fruit and nuts, 106
  Nori crêpes, 90
  Pear jello, 170

Pecan pie, 110
Pumpkin pie, 110
Tapioca pudding, 89
Dips
  Greek-style potato spread, 121
  Skordalia, 121
Dredging, 53
Dressings
  Agar strands, 106
  Creamy wakame dressing, 73
  Dulse, 120
  Korean sesame, 68
  Lemon and oil, 81
  Nut butter, 83
  Red wine vinegar and oil, 49
  Sesame oil, 49
  Vinaigrette, 49
  Vinegar miso, 94
Dulse
  Chowder, 119
  Cream of, 119
  Dressing, 120
  Hash, 123
  Kale and, 124
  Mashed potatoes with, 123
  Nut milk soup with, 118
  Pickled celery and, 125
  Potato salad and, 120
  Potato soup and, 119
  Pudding, 126
  Sauteed, 121
  Simmered, 118
  Sprouts sandwich and, 121
  Zucchini and, 124
Dulse, used in
  Casserole, 123
  Green salad, 121
  Mediterranean salad, 120
  Okra dulse deluxe, 124
  Party favors, 126
  Potato spread, 121
  Vegetable pie, 122
Dulse chowder, 119
Dulse dressing, 120
Dulse hash, 123
Dulse and kale, 124
Dulse & potato salad, 120
Dulse pudding, 126
Dulse simmered in broth, 118
Dulse & sprouts sandwich, 121
Dulse and zucchini cornucopia, 124

Eggplant, Mediterranean salad with, 146
Empanadas, Hijiki, 97
Entrée
　Arame and Carrots, 58
　Baked Stuffed Pumpkin, 98
　Brussels Sprouts & Arame Teriyaki, 47
　China, Japan in a Wok, 83
　Color Show, 96
　Dulse and Zucchini Cornucopia, 124
　Hijiki rice, 95
　Hijiki with black beans, 92
　Kidney beans, snap beans and arame, 60
　Lentil spread, 74
　Mexican fiesta, 97
　Nori rolls, 85
　Nori tofu, 84
　North African casserole, 123
　Scrambled tofu and arame, 56
　Shepherd's pie, 98
　Soba noodles, 68
　Soft rice, 72
　Soybeans with kombu, 148
　Soybeans and wakame stew, 69
　Sprouts and wakame, 74
　Stir-fried arame, 60
　Tahini quiche, 99
　Tamale pie, 57
　Tempura, 88
　Tofu with hijiki, 96
　Vegetable pie, 122
　Vegetable stew, 71

Filling, empanadas, with hijiki, 97
Filling, nori-maki
　Nori rolls, 30
　Nori *tsukudani*, 85
　Party favors, 126
　Umeboshi, 89
Filling, sweet miso-tahini, in nori crepes, 90
"Foggy mountain" soup with snow peas, 65
Fresh sprout salad with nori, 81
Fried tofu, spinach, and nori salad, 81
Fruits, used in (*see also specific fruits*)
　Citrus jam, 111
　Grape jello cubes, 109
　Mediterranean salad, 120
　Molded fruits and nuts, 106

Garlic and kombu pickles, 52
Garlicky pumpkin seeds, 43
Grape jello cubes with fresh fruit compote, 109
Grapefruit dessert, 109

Greek-style potato spread (*skordalia*), 121
Greens, used with
　Dulse and kale, 124
　Dulse and potato salad, 120
Greens & sea greens, 121
Greens topped with sesame seeds and nori (*o-hitashi*), 87

Harusame, see Noodles, clear
Hash, dulse, 123
Henry's favorite, 181
Hijiki
　black beans with, 92
　broccoli and, in soup, 92
　carrots and, 100
　color show using, 96
　curried, with apple salad, 93
　empanadas using, 97
　Mexican fiesta using, 97
　mock crabmeat using, 100
　potato salad, 93
　rice, 95
　sesame seeds with, 95
　shepherd's pie using, 98
　sweet potato and, 99
　tahini quiche using, 99
　tofu with, 96
　tomato with miso using, 94
　　wakame salad and, 94
　stuffed pumpkin using, 98
Hijiki with black beans, 92
Hijiki and carrots with creamy "white sauce", 100
Hijiki empanadas, 97
Hijiki potato salad, 93
Hijiki and wakame salad, 94
Hot potatoes and kombu, 49
Hot sauce with umeboshi and nori, 88
Hot & sour arame soup, 55

Irish moss
　Cranberry dessert sauce using, 116
　Raw roots salad using, 115
　Tomato aspic, 115
　Wine jelly using, 116
Irish moss tomato aspic, 115

Jams
　Citrus, 111
　Strawberry, 111
Japanese agar tofu (*takigawa-dofu*), 77
Japanese aspic with ginger sauce, 105
Japanese 7-spice seasoning (*shichimi-togarashi*), 77
Japanese-style soybeans with kombu, 48
Japanese vegetable stew (noppei), 70

Jello
　Grapefruit dessert using, 109
　Cubes with fresh fruit compote using, 109
Jelly, wine, 116
Jerusalem artichokes (sunchokes)
　Kombu pickles and, 51
　Raw roots salad using, 115
　Scalloped, 72
Juice
　Apple cider, baked spiced pears in, 107
　Grape, jello cubes using, 109
　Tomato, aspic and Irish moss, 115

Kale, dulse and, 124
Kelp powder
　Almonds using, 43
　Pumpkin seeds using, 43
　Vegetable bouillon in, 43
　Zucchini using, 43
Kelp powder vegetable bouillon, 43
Kidney beans, snap beans, and arame in tomato sauce, 60
Kombu
　Avocado and, in soup, 44
　Avocado split using, 46
　Black-eyed peas using, 48
　Bouillon, 44
　Bouillon aspic using, 103
　Candied, 52
　Chinese cucumber salad using, 45
　Condiment, 50
　Cucumbers and, pickled, 52
　Deep-fried, 53
　Garlic and, pickles, 52
　Korean cucumber salad using, 31
　Mediterranean salad with, 46
　Nori rolls using, 85
　One-pot sushi using, 86
　"Pickled" in ginger and soy, 51
　Potatoes and, 49
　Rice using, 71
　Soba noodles using, 68
　Soybeans with, 48
　Sunchokes and, pickles, 51
　Sweet potato and, 49
　Sweet and spicy, 52
　Tomato sauce, in, 49
　Tsukudani, 50
　Wrapped in tofu, 47
Kombu bouillon (*dashi*), 44
Kombu in continental tomato sauce, 49
Kombu pickled in ginger and soy, 51
Kombu and sweet potato, 49
Kombu *tsukudani*, 50
Kombu-wrapped tofu, 47

Korean cucumber salad, 45
Korean salad, 68
Korean-style nori, 77
Kudzu
　China, Japan in a wok using, 83
　Dulse pudding using, 126
　Hot and sour arame soup using, 55
　Scalloped sunchokes using, 72

Layered avocado & vegetable mold, 104
Leek, cream of dulse and potato soup using, 119
Lemon, wine jelly using, 116
Lemon bouillon aspic, 103
Lentil salad Arabian-style, 82
Lentil salad Ethiopian-style, 82
Lentil spread, 74
A little China, a little Japan in a wok, 83

Marinade
　Pickled celery and dulse using, 125
　Vinegar gingered hijiki rice using, 95
Marinated agar strands and spinach salad, 106
Mashed potatoes with dulse, 123
Mediterranean salad with dulse, 120
Mediterranean salad with kombu, 46
Mexican fiesta, 97
Milk, almond, almond sweet, 108
Milk, nut
　Dulse pudding, in, 126
　Soup with dulse, 118
Miso
　Nori and, 89
　Nori crepes using, 90
　Open-face sandwich using, 66
　Soup for all seasons, 63
　Summer soup, 63
　Tomato with, 94
　Winter soup, 63
A miso soup for all seasons, 63
Mizu Yokan, 112
Molasses spice aspic, 106
Molded fruit & nuts, 106
Mushrooms
　China, Japan in a wok using, 83
　Color show using, 96
　Dulse hash using, 132
　Dulse and sprouts sandwich using, 121
　Dulse and zucchini cornucopia using, 124
　Greens and sea greens using, 121
　Nori *tsukudani* using, 85
　Okra dulse deluxe using, 124
　Shepherd's pie using, 98
　Soft rice using, 71

Wakame vinaigrette using, 67
Noodles, clear (*Harusame*)
　Cucumber, wakame and, salad, 67
Noodles, soba
　Arame and carrots using, 58
　with wakame, fried tofu and carrots, 68
Noppei, 70
Nori
　African peanut soup using, 78
　China, Japan in a wok using, 83
　Crepes, 90
　Cucumber and grated radish, and, 80
　Greens and sesame seeds and, 87
　Henry's favorite using, 81
　Hot sauce with umeboshi and, 88
　Japanese 7-spice seasoning using, 77
　Korean-style, 77
　Lentil salad Arabian-style using, 82
　Lentil salad Ethiopian-style using, 82
　Miso "pickle" and, 89
　Nuts and, 78
　On the side, 87
　One-pot sushi using, 86
　Oriental cole slaw using, 80
　Party favors, 126
　Rolls, 85
　Soba noodles using, 68
　Soft rice using, 71
　Sprout salad with, 81
　Sweet potato and hijiki balls using, 99
　Tapioca pudding with, 89
　Toasted sheets, 76
　Tofu, nut butter and, salad, 83
　Tsimmas with, 86
　Tempura using, 88
　Soup with umeboshi, 78
　Tofu with rice, 84
　Tsukudani, 85
　Umeboshi and, condiment, 89
　Umeboshi and watercress and, salad, 79
　Wontons using, 79
　Wrapped fried tofu, 84
Nori crêpes, 90
Nori and cucumber with grated radish, 80
Nori and miso, 89
Nori-miso "pickle", 89
Nori rolls (*nori-maki*), 85
Nori on the side, 87
Nori soup with umeboshi, 78
Nori-tofu with rice, 84
Nori *tsukudani*, 85
North African casserole (*tagine*)
Nuts, Almonds
　Nuts and Nori, 78
　Tangy, 43

133

Nuts, mixed, tsimmas with nori, 86
Nuts, pecans, pecan pie, 110
Nuts, walnuts, pie, 110
Nut milk soup with dulse, 118
Nuts & nori Hawaiian-style, 78

Oats, sea vegetable croquettes using, 125
O-hitashi, 87
Okra dulse deluxe, 124
O-kayu, 71
One-Pot-Dish
  Soba noodles, 68
  Soft rice, 71
  Sushi, 86
  Vegetable stew, 70
One-pot sushi (*chirashi-zushi*), 86
Onions, using
  Dulse hash, 123
  North African casserole, 123
  Shepherd's pie, 98
Orange, dulse pudding using, 126
Orange, Mandarin, almond sweet using, 108
Oriental cole slaw, 80

Parsnips
  Winter miso soup using, 63
  Zen tempura using, 63
Party favors, 126
Pear
  Baked spiced, in cider aspic, 107
  Pureed jam, 107
Pecan pie, 110
Pepper, bell
  Baked stuffed pumpkin using, 98
  Dulse and zucchini cornucopia using, 124
Pepper, green
  Sweet potato and hijiki balls using, 99
  Wakame succotash using, 72
Pepper, hot, Mexican fiesta using, 97
Pickled celery and dulse, 125
Pie
  Pecan, 110
  Pumpkin, 110
  Shepherd's, 98, 122
  Tahini quiche, 99
  Vegetable, 122
Pie, crust, tahini quiche, 68
Pickle
  Celery and dulse, 125
  Cucumber and kombu, 52
  Garlic kombu, 52
  Kombu ginger-soy, 51
  Nori-miso, 89
  Sunchoke and kombu, 51

Sweet and spicy, 52
Pickle, boiled, nori *tsukudani*, 85
Potato
  Broccoli salad using, 56
  Dulse hash using, 123
  Greek-style spread, 121
  Hot with kombu, 49
  Mashed with dulse, 123
  Salad, with dulse, 120
  Salad, with hijiki, 93
  Soup, cream of dulse and, 119
  Tahini wakame soup using, 64
  Tsimmas with nori using, 86
  Vegetable stew using, 71
  Winter miso soup using, 63
  Zen tempura using, 88
Potato, sweet
  Hijiki balls, and, 99
  With kombu, 49
  Mock crabmeat using, 100
Pudding, tapioca with nori, 89
Pumpkin, baked stuffed, 98
Pumpkin pie, 110
Pureed pear jello, 107

Quiche, tahini, 99

Radish, daikon
  Nori and cucumber, with, 80
  Nori soup using, 78
  Umeboshi, nori, and watercress salad using, 79
Raw roots salad, 115
Rice
  Baked stuffed pumpkin using, 98
  Nori rolls using, 85
  Nori tofu with, 84
  North African casserole using, 123
  One-pot sushi using, 86
  Soft, 71
  Sweet vinegar gingered, 95
Rice, accompaniment to
  Arame and cabbage, 59
  Cucumbers and kombu pickled, 52
  Garlic and kombu pickles, 52
  Henry's favoite, 81
  Kombu in tomato sauce, 49
  Kombu *tsukudani*, 50
  Kombu-wrapped tofu, 47
  Nori-miso "pickle," 89
  Nori-tofu, 84
  Nori *tsukudani*, 85
  Nori-wrapped fried tofu, 84
  Party favors, 126
  Sautéed arame, 56
  Spicy kombu condiment, 50

Toasted nori sheets, 76
Umeboshi and nori condiment, 89
Wakame condiment, 72
Roots, beet, raw roots salad using, 115
Salad
  Agar tofu, 104
  Arame cabbage, 59
  Aspic with ginger sauce, 105
  Avocado split, 46
  Avocado and vegetable mold, 104
  Bamboo shoots and wakame, 66
  Broccoli, 56
  Chinese cucumber, 45
  Bouillon aspic, 103
  Cole slaw, 80
  Curried hijiki and apple, 93
  Dulse dressing, 120
  Greens and sea greens, 121
  Greens with sesame seeds and nori, 87
  Hijiki and wakame, 94
  Irish moss tomato aspic, 115
  Korean, 68
  Korean cucumber, 45
  Lemon bouillon aspic, 103
  Lentil Arabian-style, 82
  Lentil Ethiopian-style, 82
  Marinated agar strands and spinach, 106
  Mediterranean, with dulse, 120
  Mediterranean, with kombu, 46
  Molded fruits and nuts, 106
  Nori and cucumber, 80
  Potato, with hijiki, 92
  Pureed pear jello, 107
  Raw roots, 115
  Sautéed arame, 56
  Snow peas, 73
  Sprouts with nori, 81
  Tofu, nori and nut butter, 83
  Tomato with vinegared miso, 94
  Umeboshi, nori and watercress, 79
  Wakame vinaigrette, 67
Sandwich
  Dulse and sprouts, 121
  Wakame open-face, 66
Sauce
  cranberry dessert, 116
  dipping, 84, 88
  ginger, 105
  hot, 88
  mustard, 126
  sesame oil, 81
  soy-mustard, 104
  Szechuan hot, 96
  "white", 100
Sautéed arame with soy sauce and vinegar, 56

Sautéed dulse, 121
Sautéed hijiki with sesame seeds, 95
Savory vegetable pie with dulse, 122
Scalloped sunchokes & wakame, 72
Scrambled tofu and arame, 56
Sea vegetable croquettes, 125
Seasoning, Japanese 7-spice seasoning (*shichimi-togarashi*), 77
Seeds
  Pumpkin, 43
  Sunflower seed butter, 120
Shepherd's pie, 98, 122
Shichimi-Togarashi, 77
Side dish
  Avocado split, 46
  Avocado and vegetable mold, 104
  Arame and cabbage in mustard sauce, 59
  Baked stuffed pumpkin, 98
  Black-eyed peas with kombu, 48
  Brussels sprouts and arame teriyaki, 57
  Bouillon aspic, 103
  Chinese cucumber salad, 45
  Curried arame, 58
  Dulse hash, 123
  Dulse and kale, 124
  Dulse and potato salad, 120
  Dulse and zucchini cornucopia, 124
  Greens topped with sesame seeds and nori, 87
  Hot potatoes and kombu, 49
  Hijiki and carrots, 100
  Irish moss tomato aspic, 115
  Kidney beans, snap beans and arame in tomato sauce, 60
  Kombu in tomato sauce, 49
  Kombu and sweet potatoes, 49
  Mashed potatoes with dulse, 123
  Nori crêpes, 90
  Nori on the side, 87
  Nori and cucumber, 80
  Nori rolls, 11
  Okra dulse deluxe, 124
  Sautéed arame, 56
  Sweet cabbage and arame, 59
  Vegetable pie with dulse, 122
  Sushi, 86
  Sweet potato and hijiki balls, 99
  Tofu, nori and nut butter salad, 83
  Tomato with vinegared miso, 94
  Tsimmas, 86
Snacks
  Aspic, molasses spice, 106
  Kombu chips, 53
  Lentil spread, 74
  Mizu yokan, 112
  Molded fruits and nuts, 106

Nori crepes, 90
Nori, Korean-style, 77
Nori rolls, 85
Nori, toasted, 76
Nori-wrapped tofu, 84
Nuts and nori, 78
Party favors, 126
Sweet potato and hijiki, 99
Tapioca pudding, 89
Wakame, 66
Snow peas with creamy wakame dressing, 73
Soba noodles with wakame, fried tofu, and carrots, 68
Soft rice (*o-kayu*), 71
Soup
  African peanut, 78
  Broccoli and hijiki, 92
  Clear wakame, 62
  Cream of dulse and potato, 119
  Creamy avocado and kombu, 44
  Dulse in broth, 118
  "Foggy Mountain", 65
  Hot and sour arame, 55
  Kelp powder vegetable bouillon, 43
  Kombu bouillon, 44
  Miso, four seasons, 63
  Miso, summer, 63
  Miso, winter, 63
  Nori with umeboshi, 78
  Nut milk with dulse, 118
  Split pea minestrone, 64
  Tahini wakame, 64
  Wakame and chrysanthemum, 65
  Wonton, 79
Soybeans, *see beans, soy*
Soybeans and wakame stew, 69
Spicy kombu condiment, 50
Spicy Szechuan tofu with hijiki, 96
Split pea minestrone, 64
Spinach, used in
  Dulse and zucchini cornucopia, 124
  Salad, agar strands and, 106
  Salad, Korean, 68
  Salad, tofu and nori, 81
  Tahini quiche, 99
Spread, sandwich, 66, 73
Sprouts
  Arame, tofu and carrots with, 60
  Dulse sandwich with, 121
  Dulse and zucchini with, 124
  Oriental cole slaw with, 80
  Salad with nori, 81
Sprouts, alfalfa, 94
Sprouts, mung bean
  Avocado with, 46

Wakame and, 74
    Korean salad with, 68
Squash and Zucchini
    Mexican dish, 97
    Pumpkin pie, 110
    Salad using, 46
    Shepherd's pie, 98
    Soup using, 63
    Vegetable pie using, 122
    Zucchini, 43
Stew
    Japanese vegetable, 70
    Lentil spread, 74
    Soybeans and wakame, 69
    Vegetable, 71
Stir-fry
    Arame and carrots, 58
    China and Japan in a wok, 83
    Curried arame, 58
    Mock crabmeat, 100
    Sautéed arame, 56
    Sprouts and wakame, 74
    Wakame and bamboo shoots, 73
Stir-fried arame, tofu, carrot and bean sprouts with tahini-miso sauce
Strawberry-agar topping
Strawberry jam
Succotash, 72
Summer miso soup with wakame, 63
Sunchokes, see Jerusalem artichokes
Sunchoke & kombu pickles, 51
Sushi, one-pot, 86
Sweet cabbage and arame, 59
Sweet potato and hijiki balls, 99
Sweet & spicy kombu, 52
Sweet vinegar-ginger and hijiki rice, 95

Tagine, 123
Tahini, used in
    China and Japan in a wok, 83
    Nori crêpes, 90
    Party favors, 126
    Snow peas with dressing, 73
    Tahini quiche, 99
    Tahini and wakame soup, 64
    Tofu-nori salad, 83
Tahini quiche, 99
Tahini and wakame soup, 64
Takigawa-dofu, 104
Tamale pie, 57
Tangy almonds, 43
Tapioca pudding with nori, 89
Tempura, Zen, 88
Teriyaki, brussel sprouts and arame, 57
Toasted nori sheets, 76
Tofu
    Arame with, 56
    Arame and carrots with, 58
    Arame soup with, 55
    China and Japan in a wok using, 83
    Clear soup with, 62
    Color show using, 96
    Dressing, 73
    Dulse and sprouts sandwich with, 121
    Dulse and zucchini with, 124
    Hijiki and carrots with, 100
    Hijiki rice with, 95
    Japanese agar, 104
    Kombu-wrapped, 47
    Nori and nut butter salad with, 83
    Nori and rice with, 84
    Nori-miso "pickle" with, 89
    Nori-wrapped fried, 84
    Open-face sandwich with, 66
    Sauce, creamy, 100
    Snow peas with, 73
    Soba noodles with, 81
    Spicy with hijiki, 96
    Stir-fried arame with, 60
    Sweet cabbage and arame with, 55
    Hijiki and carrots with, 100
Tofu fruit pies, 111
Tofu, nori, and nut butter salad, 83
Tomatoes
    Beans and arame with, 60
    Black-eyed peas Texas-style with, 48
    Lentil spread, 74
    Mexican fiesta, 97
    Tamale pie, 57
Tomato "flowers" with vinegared miso, 94
Tomato sauce, kombu in, 49
Tortilla accompaniments
    Lentil spread, 74
    Mexican fiesta, 97
Tsimmas with nori, 86
Tsukudani, kombu, 50
Tsukudani, nori-mushroom, 85
Umeboshi plums
    Dulse dressing with, 120
    Hot sauce and nori with, 88
    Nori soup with, 78
Umeboshi and nori condiment, 89
Umeboshi, nori, and watercress salad, 79
Vegetable dishes
    Henry's favorite, 81
    Hijiki and carrots, 100
    Japanese vegetable stew, 70
    Nori-wrapped tofu, 84
    Scalloped sunchokes and wakame, 72
    Snow peas, 73
    Tempura, 88
    Tofu, spinach and nori salad, 81
    Tsimmas with nori, 86
    Vegetarian mock crabmeat, 100
    Vegetable stew, 71
    Wakame and bamboo shoots, 73
    Wakame succotash, 54
Vegetable stew, 71
Vegetarian stir-fried mock crab meat, 100
Vinegar, rice, 30
Wakame
    Bamboo shoots with miso and, 73
    Braised sprouts and, 34
    Clear noodle salad and, 67
    Chrysanthemum soup and, 65
    Clear soup with, 62
    Condiment
    Creamy dressing, 73
    Open-face sandwich, 66
    Salad with hijiki, 94
    Scalloped sunchokes and, 72
    Sea salad and bamboo shoots, 66
    Soba noodles and tofu with, 68
    Stew and soybeans, 69
    Succotash, 72
    Summer miso soup with, 63
    Tahini soup and, 64
    Vinaigrette, 67
    Winter miso soup with, 63
Wakame, used in
    "Foggy Mountain" soup, 65
    Japanese aspic, 105
    Japanese vegetable stew, 70
    Korean salad, 34
    Lentil spread, 34
    Miso soup, 63
    Soft rice, 71
    Split pea minestrone, 64
    Vegetable stew, 71
    Zen tempura, 88
Wakame and bamboo shoots with miso, 73
Wakame and chrysanthemum soup, 65
Wakame condiment, 72
Wakame open-face sandwich, 66
Wakame succotash, 72
Wakame vinaigrette with cauliflower and string beans, 67
Watercress
    Avocado and vegetable mold, 104
    Umeboshi, nori and, salad, 79
Wine jelly, 116
Winter miso soup with wakame, 63
Wok dish
    Arame, 56

Arame and carrots, 58
Arame, tofu, carrots and bean sprouts, 60
Chinese cucumber salad, 45
Curried arame, 58
Dulse and zucchini cornucopia, 124
Hijiki with sesame seeds, 95
A little China, a little Japan, 83
Okra dulse deluxe, 124
Spicy Szechuan tofu with hijiki, 96
Tempura, 88
Wakame and bamboo shoots, 73
Wontons as you want, 79

Yam
    African peanut soup using, 78
    Shepherd's pie using, 98
    Tsimmas with nori using, 86

Zen tempura, 88
Zucchini, *see Squash and Zucchini*
Zucchini a l'Italienne, 43

135

ABOUT THE AUTHOR

Born in Washington, D.C., in 1944, Sharon Ann Rhoads holds a Masters Degree in Oriental studies from the University of Michigan. Fluent in Japanese, she spent eight years living and working in Japan studying that nation's traditional culture and cuisine. For five years a student of Shinzan Kamijo, one of Japan's leading contemporary masters of calligraphy, Ms. Rhoads is now an accomplished practitioner of the art. She presently lives in New York City, where she is pursuing a career as an artist, writer, and teacher of natural-foods cookery.